byas

Film Formation

ACS SYMPOSIUM SERIES **941**

Film Formation

Process and Morphology

Theodore Provder, Editor
Eastern Michigan University

Sponsored by the
ACS Division of Polymeric Materials: Science and
Engineering, Inc.

American Chemical Society, Washington, DC

Library of Congress Cataloging-in-Publication Data

Film formation : process and morphology / Theodore Provder, editor.

    p. cm.—(ACS symposium series ; 941)

    Sponsored by the ACS Division of Polymeric Materials: Science and Engineering, Inc."

    Includes bibliographical references and index.

    ISBN 13: 978–0–8412–3961–6 (alk. paper)

    ISBN 10: 0–8412–3961–4

    1. Protective coatings—Congresses. 2. Surfaces (Technology)—Congresses. 3. Emulsion paint—Congresses.

    I. Provder, Theodore, 1939- II. Series.

TA418.76.F537   2006
667′.9—dc22
                                              2006042789

The paper used in this publication meets the minimum requirements of American National Standard for Information Sciences—Permanence of Paper for Printed Library Materials, ANSI Z39.48–1984.

PRINTED IN THE UNITED STATES OF AMERICA

Foreword

The ACS Symposium Series was first published in 1974 to provide a mechanism for publishing symposia quickly in book form. The purpose of the series is to publish timely, comprehensive books developed from ACS sponsored symposia based on current scientific research. Occasionally, books are developed from symposia sponsored by other organizations when the topic is of keen interest to the chemistry audience.

Before agreeing to publish a book, the proposed table of contents is reviewed for appropriate and comprehensive coverage and for interest to the audience. Some papers may be excluded to better focus the book; others may be added to provide comprehensiveness. When appropriate, overview or introductory chapters are added. Drafts of chapters are peer-reviewed prior to final acceptance or rejection, and manuscripts are prepared in camera-ready format.

As a rule, only original research papers and original review papers are included in the volumes. Verbatim reproductions of previously published papers are not accepted.

ACS Books Department

Contents

vii

Indexes

Preface

Coatings technologies continue to be influenced by the need to lower volatile organic contents (VOC) in order to comply with stricter environmental regulations as well as to reduce the use of costly petroleum-based solvents. During the past one to two decades, the use of waterborne coatings in the architectural, industrial maintenance, and original equipment manufacturing (OEM) sectors has continued to grow replacing solvent-based coatings while meeting the ever decreasing VOC targets. In addition to waterborne coatings, other alternative technologies in the industrial and OEM sectors include powder coatings, UV-curable coatings, and high-solids coatings. Understanding the film formation process relevant to these technologies from both a scientific and technological perspective continues to be of prime importance for achieving quality products that can effectively compete in the marketplace.

The previous two books (*1, 2*) .in this series demonstrated the vitality of this field of investigation. The increasingly competitive global marketplace and the changing technical and economic environment amplifies the current and continuing need for generating knowledge about the film formation process, which scientists and technologists can then apply to developing quality coatings products. The growth of nanomaterials technology has impacted film formation science and technology and has enabled the development of novel morphologies and film structures. Unique and new characterization methodology continues to be applied to the understanding of the film formation process.

The first section of this book focuses on kinetics and mechanism of film formation. Knowledge of the kinetics involved in the film formation process provides insights to understanding the mechanism of the process. In this section, the kinetics of UV-curable systems are investigated and a better understanding of the effect of surfactants on film formation, morphology, and film structure in waterborne systems is gained. The influence of nanomaterials technology on the formation of unique morphologies and film structures is very evident.

I expect that this book will challenge and encourage scientific and technological investigators to continue expanding the knowledge in this field as well as applying the knowledge to commercially relevant coatings systems. I thank the authors for their effective oral and written communications and the reviewers for their helpful critiques and constructive comments

References

1. *Film Formation in Waterborne Coatings;* Provder; T., Winnik, M. A.; Urban, M. W., Eds.: ACS Symposium Series 648; American Chemical Society, Washington, DC, 1996.
2. *Film Formation in Coatings: Mechanisms, Properties, and Morphology;* Provder, T.; Urban, M. W., Eds.: ACS Symposium Series 790; American Chemical Society, Washington, DC, 2001.

Theodore Provder, Director
Coatings Research Institute
Eastern Michigan University
430 West Forest Avenue
Ypsilanti, MI. 48197

Film Formation

Kinetics and Mechanism

Chapter 1

UV-Induced Film Formation Kinetics as a Function of Film Thickness and in the Adjacent Dark Non-Irradiated Region

Yuemei Zhang, Justin Doo, David Krouse, and David Kranbuehl[*]

Departments of Chemistry and Applied Science, College of William and Mary, Williamsburg, VA 23187
[*]Corresponding author: dekran@wm.edu

The kinetics of UV cure and thermal cure have been monitored using modulated differential scanning calorimetry (DSC) and by frequency dependent dielectric sensing (FDEMS) using a planar interdigitated sensor. The FDEMS technique has the ability to monitor the film formation process at a particular depth in a coating while the DSC measures the averaged rate of cure through the entire thickness. This report focuses on the use of FDEMS sensors to monitor the effect of oxygen diffusion on the cure of an acrylic resin, on the effect of film thickness on the cure of a coating by UV radiation and on the effect of the line width in a mask on cure in the "masked" non-irradiated region during UV cure. The results show that the rate of UV cure varies with the depth. The effect is particularly true during thermally initiated cure in the presence of oxygen. The results also show that when using a mask with a UV cure resin, the rate of cure decreases with increasing width of the masked lines in the mask's non-irradiated regions.

3

Introduction

The ability to monitor the cure of a film is particularly important for the coatings industry. Most measurements of a film as it cures utilize conventional laboratory techniques such as modulated differential scanning calorimeter (DSC), dynamic mechanical measurements (DMA), spectroscopic techniques involving infrared, visible or UV absorption and simple physical tests, such as "dry to touch" and "dry to hard". The problem with DSC and DMA measurements is that it is difficult to monitor the cure process as a thin film with one side of the film in contact with the surface to be coated and one side exposed to the environment under the application conditions. Spectroscopic techniques overcome this problem. However, in general they lack sensitivity during the achievement of full cure. Monitoring the final build up to complete cure is particularly important when monitoring and verifying durability. Physical criteria, such as "dry to hard" are subject to considerable variation from one person to another.

On the other hand, frequency dependent dielectric measurements using in situ micro sensors, FDEMS, is a particularly useful technique for monitoring the changing state of a coating during synthesis, cure and aging. Measurements can be made in the laboratory to monitor the polymerization process in a flask, to monitor cure as a coating in an oven or under a UV lamp, and to monitor the coating's durability and aging in a weather controlled environmental chamber or other degrading environment. Thus the FDEMS micro sensor technique ought to be more widely used, particularly in monitoring cure and degradation of a coating. An important reason the planar micro sensor should be used more extensively is that it is ideally suited to monitoring coatings where only one side of a thin polymer film is exposed to the environment, a condition which is difficult to duplicate in most other measurements.

A review of the FDEMS technique has been recently published (1). Other articles describe its application to a variety of monitoring needs in the coating industry (2-11). This article reports on the use of FDEMS sensors to monitor the effect of oxygen diffusion on the cure of an acrylic resin, on the effect of film thickness on the cure of a coating by UV radiation and on the effect of the line width in a mask on cure in the "masked" non irradiated region during UV cure.

Experimental

Materials

Two systems were studied in this paper. The first system is cured using UV radiation. It is composed of 80wt% dimethacrylate of tetraethoxylated bisphenol A (D121, Akzo-Nobel), 20wt% styrene (ST, Aldrich), 0.5 phr of chain transfer agent 1-dodecanethiol (Aldrich) and 1 phr of initiator Irgacure819 (Ciba). The second system is cured thermally. It is composed of Isobornyl methacrylate(IBoMA, Aldrich, with 150 ppm methylether hydroquinone as inhibitor) and 2 phr of initiator Benzoyl peroxide(BPO, Akzo-Nobel). All products were used as received.

Equipment

Dielectric Impedance Measurements

A HP 4192A LF impedance analyzer is used to monitor the thermal cure of IBoMA. A HP 4263A LCR meter is used to monitor the cure using UV radiation. The HP impedance analyzer and LCR meter were connected to a planar interdigitated kapton dielectric sensor with an air replaceable capacitance C_o. The sensor was used to monitor the changes in conductance and thereby the dielectric permittivity's loss ε'' of the sample during the polymerization process over a range of frequencies from 0.1KHz to 100KHz. The dielectric loss ε'' was calculated for each frequency, f, where $\varepsilon'' = G/\omega C_o$ and $\omega = 2 \pi f$. The sensor was placed on the bottom of a mold. The mold was filled with liquid placed on the top of the sensor. The planer sensor monitors changes on its surface. Since the sensor is on the bottom of the mold, it is monitoring changes at a depth equal to the thickness of the film. The depth of the mold determines the thickness of the sample.

DSC Measurement

The thermal polymerization kinetics and UV cure kinetics were also studied by DSC using a TA Instruments 2920 modulated DSC to compare and correlate these results with the dielectric data. The thermal analysis consisted of an isothermal DSC run to determine ΔH_1 at each temperature. Then this was followed by a temperature ramp to a suitable high temperature to determine the residual heat (ΔH_2) and thereby the total heat of the polymerization ($\Delta H_1 + \Delta H_2$). The cure extent at a given time was calculated from: $\alpha(t) = \Delta H(t)/(\Delta H_1 + \Delta H_2)$, where $\alpha(t)$ is the cure extent and $\Delta H(t)$ is the heat produced by the reaction at time t.

UV-Radiation source

A 200 watt Hg arc lamp (Oriel Instrument) equipped with a fiber optic cable was used to guide the light to the sample chamber in the DSC cell and on to the dielectric mold assembly sensor. The spectral output of the lamp was controlled with a 365nm-365nm band pass filter with a peak irradiation wavelength of 361 nm.

Results and Discussions

UV-Cure System

The effect of thickness on UV cure

Dielectric measurements were used to monitoring the UV cure at room temperature. Using the Debye equation for a single dipolar relaxation time the frequency dependence of the loss parameter ε'' can be expressed as

$$\omega\varepsilon''(\omega) = \frac{\sigma}{8.85*10^{-14}} + (\varepsilon_0 - \varepsilon_\infty)\frac{\omega^2\tau}{1+(\omega\tau)^2} \tag{1}$$

The dielectric loss factor ε" is a summation of two factors. The first factor is contributed by the ionic mobility and the second factor is contributed by the rotational mobility of diploes. When the frequency is low and at low cure extent, the contribution of the dipolar rotational mobility is very small. Therefore at low frequencies, equation (1) can be written as equation (2) where

$$\omega\varepsilon''(\omega) = \frac{\sigma}{8.85*10^{-14}} \tag{2}$$

σ is the specific conductivity in units of ohm^{-1}cm^{-1} and the permittivity of free space is $8.85*10^{-14}$C^{2}J^{-1}cm^{-1}.

A plot of log(ωε") versus time during the polymerization shows the ionic mobility contribution to the dielectric loss at low frequency, 0.1kHz. Figure 1 shows the decrease of log (ωε") as a function of depth during the film formation process. The ion mobility decreases with the extent of cure because the viscosity increases. Its changing value can be correlated with the cure extent. To study the reaction at different depths, different thicknesses of sample were applied on the top of the sensor. The reduction in rate versus the thickness of the film at the surface of the sensor is due to photons being absorbed by the initiator molecules in the upper layer. Thus the intensity of the radiation decreases with depth. The photo-cure rate can be expressed by the equation (12)

$$-\frac{d[M]}{dt} = k_p[M][R] = (\frac{fk_p^2}{k_t})^{1/2}(\Phi I_{abs})^{1/2}[M] \tag{3}$$

and $I_{abs}=I_0(1-e^{-2.3\varepsilon lc})$, (13) where ε is molar extinction coefficient(in M^{-1} cm^{-1}), l is optical path length(cm) and c is concentration of photoinitiator (in M) . I_{abs} increases with thickness (l) of the sample. So the incident light at the bottom, I_{ob}, is equal to I_0-I_{abs}, which decreases with thickness or depth. When the system and cure condition are the same, the photo-polymerization rate at the bottom is proportional to $(I_{ob})^{1/2}$. This is the reason that the film formation process is a function of depth and it is slower at the bottom of the coating. In addition, the final value of $\log(\omega\varepsilon')$ is also a function of depth. From Figure 1, at the end of the reaction, the value of $\log(\omega\varepsilon'')$ is higher for the thicker films. This indicates the final cure extent decreases with depth because of the reduction of the light intensity. The effect of light intensity on the cure extent has been studied in the paper of L. Lecamp (14), and the outcome is that the higher the intensity, the higher the final conversion.

In order to correlate the FDEMS data and the extent of reaction as measured by photo-DSC, measurements were made on a thin film at the same intensity so that there is very little effect of thickness. The correlation of $\log(\omega\varepsilon'')$ and extent of cure as determined by the DSC measurements are shown in Figure 2. The value of $\log(\omega\varepsilon'')$ changes slowly with reaction advancement initially because the resin retains a relatively fluid state. Then it drops sharply on a log scale as the film formation process occurs. This occurs at an extent of reaction of 30-45%. After that, the value of $\log(\omega\varepsilon'')$ does not change much. This indicates the polymerization is almost complete.

UV cure in the irradiated and non irradiated regions when using a mask

Cure that occurs in the dark non irradiated regions near the interface of the mask's, non irradiated dark area and the exposed area is due to the diffusion of free radicals across the boundary of the exposed area, see Figure 3.

In order to examine this and characterize cure under the mask, a series of mask experiments were designed. The total clear area and dark area are the same for a series of masks with differing widths and therefore number of dark strips. In these experiments, the total irradiated area and the masked dark area were 50% to 50%. The dark strips are the same width as the clear exposed strips and alternate across the sample of fixed total area. Examining the dielectric data of the mask experiment shown in Figure 4, the film forming process can be divided into two periods. The first period occurs from the beginning to the 10th minute. The second period occurs after the 10th minute continuing over a much longer time.

The curing during the first period follows the kinetics of the clear mask. They are similar because the intensity of the light in the clear areas are the same for masks of different strip widths. But in the second period, there are significant differences. These differences are due to variation in the amount of diffusion across the boundary between the dark and clear areas. The initiator radicals and activated monomers after being excited by the light diffuse process across this interface of the clear to the dark area and initiate reaction under the dark strips. In addition, the chains diffuse and grow into the dark non-irradiated regions. Therefore, the curing under the dark strips is affected by the mobility of these activated species, the distance from the interface and the number or total

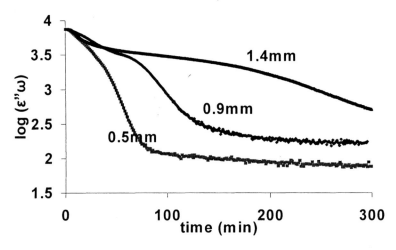

Figure 1. FDEMS data of the UV curing process at different depths at the frequency of 0.1kHz and an Intensity $I=1.5uw/cm^2$

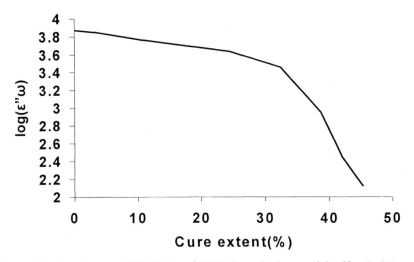

Figure 2. Correlation of FDEMS and DSC data, thickness of the film $T=0.5mm$, intensity $I=1.5uw/cm^2$, frequency=0.1kHz

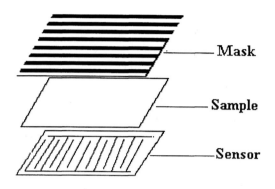

Figure 3. Mask experiment

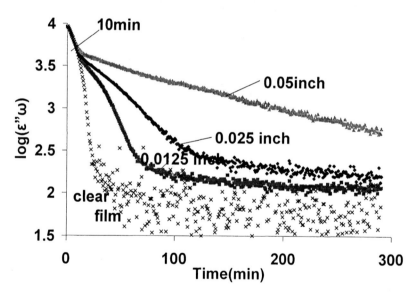

Figure 4. Mask experiment measured by FDEMS at the frequency of 0.1kHz at the depth of 0.05mm, mask width is from 0.0125inch to 0.05inch

length of the interface region. For the mask with narrower strips, the diffusion distance to reach all of the dark area is shorter and the number, or say length, of the interface region is larger. Therefore for the same condition, the narrower the strips, the faster the cure in the dark non-irradiated regions.

Another series of experiments are shown in Figure 5, which are carried out at a different depth showed a similar outcome.

In order to study the thickness effect on the cure in the non irradiated region, a series of mask experiments were conducted using masks with the same strip width but different depths, due to varying the film thicknesses, under the same intensity of light. From Figure 6, the cure rate in both clear areas and dark areas decreases with depth, at the bottom of the film. The dark cure is not only affected by the mobility of the activated molecules but also is affected by the intensity of the light and therefore the depth at which the cure is occurring.

Thermal Cure

First, the effect of oxygen on the thermal polymerization kinetics of IBoMA with 2% BPO at 80 °C was studied by DSC. Plotting the cure rate(α) versus time in Figure 7, it is seen that the polymerization without oxygen reached the highest reaction rate much earlier than the polymerization under oxygen. This is because the oxygen that diffused from the environment reacts with the free radicals and inhibits the polymerization. Since the oxygen diffuses from the surface of the sample to the bottom, its effect is a function of depth. Therefore, the polymerization kinetics is also a function of depth. It was also observed that both reactions are inhibited at the beginning of the reaction. This is because there is some inhibitor and oxygen in the IboMA system initially.

Figure 8 shows both reactions reach the final cure extent of 80%, but the polymerization without oxygen is complete after 11minutes which is much earlier than under oxygen where the completion occurs after 21minutes.

Since the kinetics of the reaction is a function of depth, it is not as interesting to study the oxygen effect on the reaction by DSC because DSC measures an average of the reaction rate with depth instead of the reaction at a particular depth. In order to study the oxygen effect in greater detail, FDEMS experiments were conducted at the same thickness as in the DSC experiment. The polymerization without oxygen is shown in Figure 9. Dipole relaxation peaks occur around the 10th to 11th minute. They reflect the build up of the glass transition temperature. The Tg peaks occur at the same time as the completion of the DSC reaction because the glass transition quenches the reaction and the reaction ceases when the cure temperature is lower than the glass transition temperature.

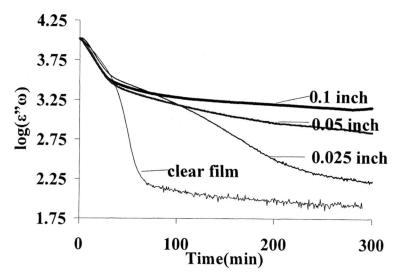

Figure 5. Mask experiment measured by FDEMS at the frequency of 0.1kHz at the depth of 0.5mm, mask width is from 0.025inch to 0.1inch

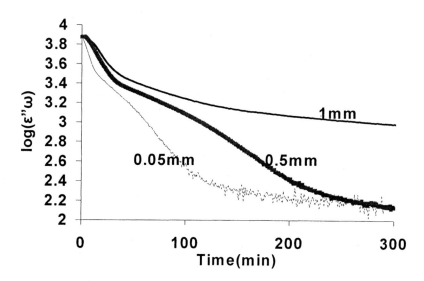

Figure 6. Mask experiments by FDEMS at the frequency of 0.1kHz, at the thickness of 0.05mm, 0.5mm and 1mm, 1=1.5uw/cm², strip width=0.025inch

For the experiment under oxygen shown in Figure 10, the dipole relaxation peak occurs around the 6th to 17th minute, which is much earlier than the end of the reaction that was measured by DSC at the 21st minute. This is because FDEMS measures the reaction process at the bottom of the film where there is much less diffusion of oxygen than has occurred at the surface. Therefore, the polymerization at the bottom is faster than the polymerization at the surface and the average rate for the bulk of the sample.

Figure 11 shows the correlation of the cure extent with the value of $\log(\omega e")$ without oxygen. From Figure 10, the relation of $\log(\omega e")$ and cure extent is almost linear up to 70% conversion. After that, the change of $\log(\omega e")$ is much larger than the cure extent that was measured by DSC. This is because FDEMS is more sensitive than DSC, especially near the end of the polymerization.

Conclusions

FDEMS can measure the polymerization rate of a film at a particular depth. The polymerization kinetics of a UV cure system is a function of depth because the light intensity decreases with depth. The cure rate at the top is faster than the bottom layer and the conversion at the top, the air film surface, is larger than at the bottom of the film at the substrate interface.

Using FDEMS to monitor UV cure under a mask, it was observed that the cure under the mask's non-irradiated regions occurs due to the mobility of activated molecules. In addition, the dark cure occurs more slowly if the width of the mask's strips increases and with depth. The rate of dark cure was determined by not only the mobility of the radicals but also the length of the exposed-masked interface.

The cure kinetics of IBoMA film under oxygen is a function of depth because of the diffusion of oxygen. The cure rate at the upper layer was less because the oxygen reacts with the radical initiator molecules.

The FDEMS planer sensor measurements were shown to be a convenient and effective way to monitor cure in a film and in particular to monitor cure as a function of depth and film thickness.

References

1. Kranbuehl, D.E., *J. Coatings Technology*, 2004, pp. 48-55.
2. Kranbuehl, D.E., *Dielectric Spectroscopy of Polymeric Materials*, 1997, pp. 303-328.
3. Kranbuehl, D.E., *Processing of Composites*, Loos, A. (Ed.), 2002, pp. 137-157.

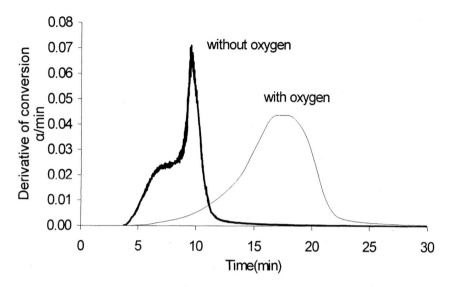

Figure 7. Using DSC to study the oxygen diffusion effect on the curing rate of IBoMA-2%BPO at 80°C

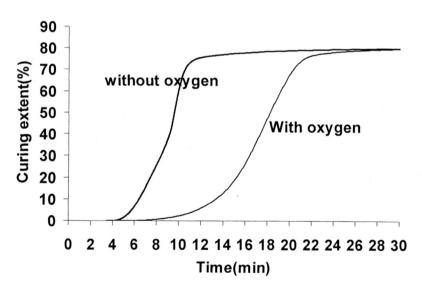

Figure 8. Using DSC to study the oxygen diffusion effect on the curing extent of IBoMA-2%BPO at 80°C

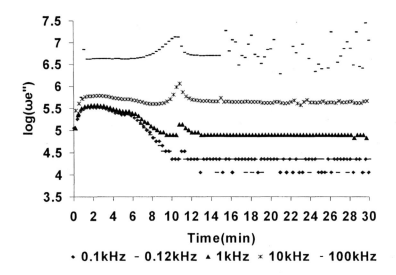

Figure 9. Using FDEMS to study the oxygen diffusion effect to the curing extent of IBoMA-2%BPO at 80°C without oxygen

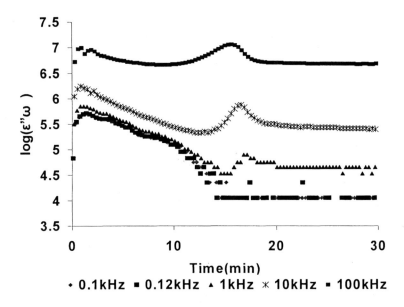

Figure 10. Using FDEMS to study the oxygen diffusion effect on the curing extent of IBoMA-2%BPO at 80°C with oxygen

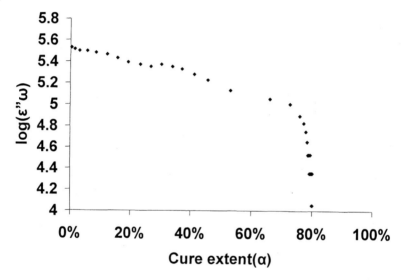

Figure 11. The Correlation of DSC and FDEMS for the cure of IBoMA+2%BPO cure at 80°C without oxygen, frequency=0.12kHz

4. Kranbuehl, D.E., Warner, J., Knowles, R. and Best, P., *60th Annual Technical Conference, Soc. Plastics Engineers*, 2002, pp. 3384-3388.
5. Kranbuehl, D. and Rogozinski, J. Polymeric Materials Science and Engineering, 81, 1999, pp. 197-198.
6. Kranbuehl, D., Hood, D. Kellam, C. and Yang, J., *Film Formation in Waterborne Coatings*, Provder, T. (Ed.), ACS Symposium Series, 1996, pp. 96-117.
7. Kranbuehl, D., Rogozinski, J. Meyer, A., and Neag, M., *Proc. 24th International Conf. in Organic Coatings: Waterborne, High Solids, Powder Coatings*, Athens, 1998, pp. 197-211.
8. Kranbuehl, D., Hood, D., Rogozinski, J. Meyer, A., and Neag, M., *Prog. Org. Coat.*, 35, 1999, pp. 101-107.
9. Kranbuehl, D., Rogozinski J., Meyer, A., Hoipkemeier, L., and Nikolic, N., *ACS Symposium Series*, 790, 2001, pp. 141-156.
10. Kranbuehl, D., Hood, D., McCullough, D., L. Aandahl, H., Haralampus, N., Newby, W. and Eriksen, M., *Progress in Durability Analysis of Composite Systems, Proc. International Conference,* Brussels, 1995, pp. 53-59.
11. Kranbuehl, D., Hood, D., Rogozinski, J., Meyer, A., Powell, E., Higgins, C., David, C., Hoipkemeier, L., Ambler, C., Elko, C., and Olukcu, N., *Recent Developments in Durability Analysis of Composite Systems*, DU-RACOSYS 99, Proc.; *4th International Conf. on Durability Anaylsis of Composite Systems*, Brussels, Belgium, 1999, pp. 413-420.
12. Flory, P., *Principles of polymer chemistry*, 1953.
13. Fouassier, J.-P., *Photoinitiation Photopolymerization and Photocuring*. 1995.
14. Lecamp, L., Youssef, B., Bunel, C., and Lebaudy, P., Polymer, 38(25) 1997, pp. 6089-6096.

Chapter 2

Copolymerization Mechanism of Photoinitiator Free Thiol–Vinyl Acrylate Systems

Tai Yeon Lee[1], T. M. Roper[1], C. Allan Guymon[2], E. Sonny Jonsson[3], and Charles E. Hoyle[1,*]

[1]Department of Polymer Science, University of Southern Mississippi, Hattiesburg, MS 39406
[2]Department of Chemical and Biochemical Engineering, University of Iowa, Iowa City, IA 52242–1527
[3]Fusion UV-Curing Systems, 910 Clopper Road, Gaithersburg, MD 20878

Abstract

The photoinitiator-free copolymerization of trifunctional thiol/vinyl acrylate mixtures as a function of thiol content has been investigated using real-time FTIR in the presence and absence of oxygen. Even without external photoinitiators, the thiol/vinyl acrylate mixtures polymerize rapidly. The addition of a multifunctional thiol to vinyl acrylate results in the conversion of the vinyl double bond of vinyl acrylate due to the preferential addition of the thiyl radical to the vinyl double bond. Both the chain transfer reaction of the carbon centered radicals to thiol and subsequent addition of the thiyl radicals to vinyl double bonds during copolymerization strongly affect the copolymerization kinetics of this system. The effect of thiol on vinyl acrylate polymerization in the presence of oxygen is particularly significant due to the role of thiol in reducing oxygen inhibition. Two free-radical chain processes occur in both the presence and absence of oxygen.

Introduction

Due to rapid growth in the field of photocuring, a significant research effort on understanding photopolymerization reactions has been expended over the past few decades.[1-5] However, the development of new technology for emerging photocuring applications is still essential in solving existing problems and improving final coating properties. New photoinitiation systems and high performance photocurable monomers have emerged to ensure expansion of the photocuring industry.

Generally, large photoinitiator concentrations are needed in photocuring formulations to reduce oxygen inhibition. One unfortunate aspect of conventional photoinitiators is that only a small fraction of the photoinitiator actually participates in the initiation process, thus leaving large concentrations of photoinitiator in the final cured films. This leads to deleterious long-term effects including leaching of photoinitiator out of the cured films, yellowing, and degradation.[1-4] Photoinitiation systems that take advantage of monomers that initiate polymerization would have many advantages. The use of such monomers has been discussed in a recent report by Hult et al.[6] in which a difunctional monomer with one maleimide reactive group and one acrylate reactive group was photopolymerized without the need of adding an external photoinitiator. Recently, Kudyakov et al.[7,8] reported that no photoinitiator is required to photoinitiate vinyl acrylate polymerization, and that the acrylate group polymerizes very rapidly while the vinyl group reacts much slower. Recent reports have dealt in greater detail with the self-initiation mechanism and polymerization kinetics of vinyl acrylate.[9,10]

The types of multifunctional monomers available for use in photocuring applications are limited compared to conventional solvent-based coatings. Recently there has been a considerable effort expended to develop new photopolymerizable monomers that lead to faster polymerization rates under ambient conditions and result in films with improved properties. In particular interest in thiol-ene monomer mixtures, which were used in industrial photopolymerization processes over two decades ago, has re-emerged.[11-14] Thiol-enes are in many ways superior to the more widely used acrylates for photopolymerization with many desirable properties including inherently rapid reaction rates, reduced oxygen inhibition, reduced problems in the final films due to film shrinkage, and good films-substrate adhesion properties.[11-14] Thiol-ene polymerization proceeds by a free-radical step growth polymerization process that is amenable to essentially any type of multifunctional ene monomer[13, 14-18] thus allowing a wide variety of crosslinked network structures to be produced.

Herein, we report on the free-radical copolymerization of trimethylol propane tris(3-mercaptopropionate)/vinyl acrylate mixtures without an external photoinitiator in the presence and absence of oxygen. Vinyl acrylate is a unique monomer with two reactive enes that have different reactivity, and as mentioned previously vinyl acrylate can self-initiate free-radical photopolymerization when exposed to light. This affords the opportunity to investigate the photopolymerization process in the absence of an added photoinitiator. Because the reactivity of thiols toward acrylate and vinyl group is totally different, we investigated the effect of thiol on vinyl acrylate polymerization as a function of thiol content. Using real-time FTIR (RTIR), we observed the polymerization of the acrylate, vinyl, and thiol groups separately. A mechanism for the polymerization of thiol/vinyl acrylate mixtures is proposed.

Experimental

Materials

Vinyl acrylate (VA), trimethylol propane tris(3-mercaptopropionate) (Trithiol), ethyl acrylate, and vinyl propionate were obtained from Aldrich Chemical Co. and used without further purification. The chemical structures of vinyl acrylate and a multifunctional thiol are shown in Figure 1 along with those of the monofunctional analogues of vinyl acrylate, ethyl acrylate and vinyl propionate.

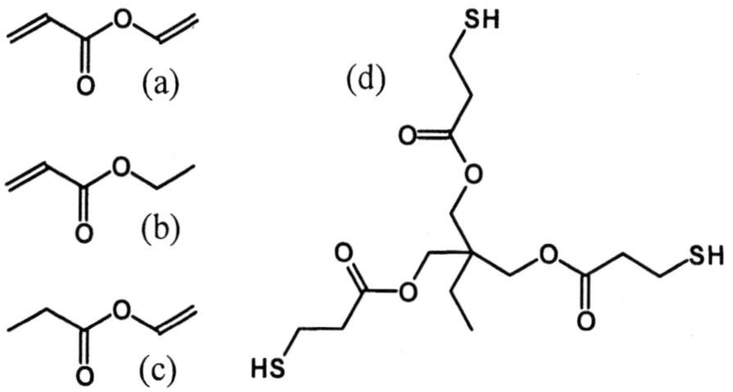

Figure 1. Chemical structures of (a) vinyl acrylate, (b) ethyl acrylate, (c) vinyl propionate, and (d) trimethylol propane tris(3-mercaptopropionate).

Methods

Real-time infrared spectra (RTIR) were recorded on a modified Bruker 88 spectrometer designed to allow light penetration to a horizontal sample using a fiber-optic cable attached to a Spectratech 200 Watt high pressure mercury xenon lamp source (ScienceTech Co.). The real-time FTIR setup has been illustrated elsewhere.[9] Samples were prepared by mixing vinyl acrylate and thiol based on the moles of each functional group. Samples were placed between two sodium chloride (NaCl) salt plates with a 15-μm Teflon spacer. To create an oxygen free atmosphere and suppress evaporation of the vinyl acrylate monomer, the edge between two salt plates was sealed by vacuum grease prior to positioning the sample on the horizontal sample holder. The polymerization kinetics in air were measured by using modified salt plates with a portion of the center carved out. The salt plate placed in the bottom had a hole 3.8 mm in diameter and 0.05 mm in depth. The salt plate covering the sample had a larger hole 12 mm in diameter and 2.0 mm in depth to allow air to enter the polymerization medium. The UV light intensity at the sample, 35 mW/cm^2, was measured by a calibrated radiometer (International Light IL-1400). The acrylate and vinyl double bonds in vinyl acrylate were monitored at 1625 cm^{-1} and 1645 cm^{-1}, respectively, and the thiol group conversion was observed at 2575 cm^{-1}. A deconvolution technique was employed to separate overlapping C=C stretching bands of the acrylate and vinyl double bonds at 1625 cm^{-1} and 1645 cm^{-1} so that the individual kinetic profiles of the acrylate and vinyl groups could be determined simultaneously.[9]

Results and Discussion

Before discussing the effects of thiol on vinyl acrylate polymerization, it is essential to understand the characteristics of vinyl acrylate homopolymerization. We have reported totally different reactivity of the acrylate and vinyl groups of vinyl acrylate.[9,10] During vinyl acrylate homopolymerization in the absence of external photoinitiators, the acrylate group polymerizes very rapidly while the vinyl group begins to react very slowly after more than 80 % of the acrylate groups have polymerized. Also, the inherent reactivity of the acrylate double bond of vinyl acrylate is much higher than traditional acrylate monomers, i.e. the polymerization rate of the acrylate group on vinyl acrylate is approximately 5-10 times greater than the polymerization rates of alkyl acrylates.[10] The enhanced reactivity of the acrylate double bond is explained by the lower electron density of the acrylate double bond compared to traditional acrylates.[10] Based on these results, concluded that vinyl acrylate produces a linear acrylate polymer followed by formation of a crosslinked network via reaction between vinyl pendant groups. Scheme I depicts a sketch of the crosslink network formation process during vinyl acrylate homopolymerization.

Because it is expected that the reactivity of thiol addition toward the acrylate and vinyl group of vinyl acrylate should be different, mixing a multifunctional thiol with vinyl acrylate homopolymerization should dramatically change the polymerization mechanism of vinyl acrylate and the resultant crosslinked network structure. Scheme II illustrates the thiol-ene polymerization mechanism. As shown, the thiol-ene reaction proceeds via a two step chain growth mechanism, a free radical addition followed by a chain transfer step.

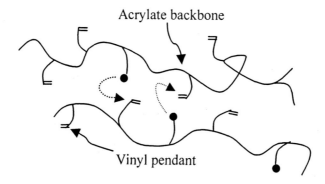

Acrylate backbone

Vinyl pendant

Scheme I. A schematic crosslink network formation process of vinyl acrylate.

The first reaction step is simply the addition of a thiyl radical to an ene double bond to generate a carbon centered radical. This radical can then undergo either a propagation reaction with another ene double bond or a chain transfer reaction resulting in the generation of another thiyl radical. The second step strongly depends on the ene radical stability and the chain transfer rate constant. For example, acrylates which form an electrophilic radical with a low chain transfer rate constant prefer homopolymerization reactions resulting in significant amounts of acrylate homopolymer. On the other hand, vinyl esters which form a relatively unstable nucleophilic carbon centered radical with a large propensity for chain transfer results in thiol-ene copolymerization. In addition, thiyl radical addition to the electron rich ene double bond of the vinyl ester group is extremely fast due to the electrophilic nature of the thiyl radical. Based on this analysis, it is reasonable to expect that the addition of the thiol to the vinyl group of vinyl acrylate would be much faster than addition to the acrylate group.

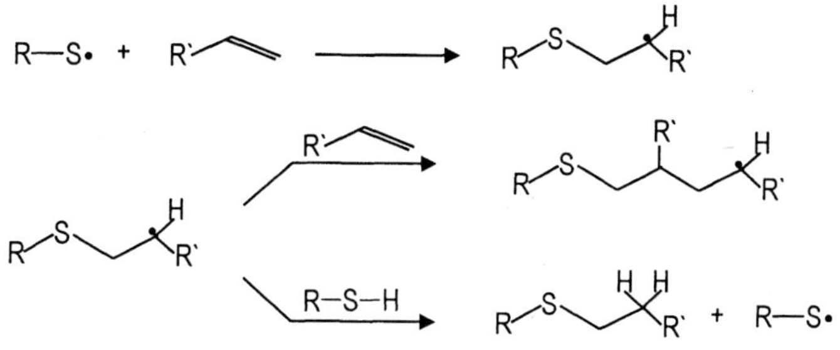

Scheme II. Thiol-ene polymerization mechanism.

22

Before attempting to investigate the polymerization of a vinyl acrylate mixture with thiol model systems consisting of thiol with each of the monofunctional analogues of vinyl acrylate, ethyl acrylate and vinyl propionate, were monitored. Figure 2 shows the conversion plots of Trithiol/vinyl propionate and Trithiol/ethyl acrylate 1:1 molar mixtures in the presence of an external photoinitiator, 2,2-dimethoxy-2-phenylacetophenone (DMPA).[10] The conversion of the Trithiol/vinyl propionate mixture reacts rapidly in a half second, much faster than the ethyl acrylate/thiol mixture. For the Trithiol/vinyl propionate mixture, thiol and vinyl group conversions are almost identical indicating an exclusive 1:1 reaction between vinyl and thiol groups, i. e., no vinyl ester radical polymerization occurs. However, for the Trithiol/ethyl acrylate mixture, 60% of the acrylate is consumed by homopolymerization while only 40% reacts with thiol. The results in Figure 2 thus leads to the conclusion that thiol reacts much faster with the electron rich vinyl propionate than the electron deficient acrylate.

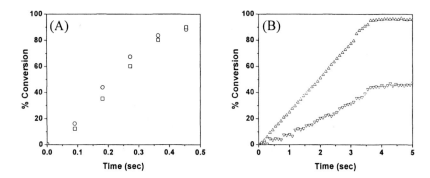

Figure 2. RTIR conversion versus time plots for the reaction of (A) (□) Trithiol/ (○) vinyl propionate and (B) (▽) Trithiol / (△) ethyl acrylate 1:1 molar mixtures upon irradiation with 0.5 wt% of DMPA using 365 nm light. Light intensity is 14 mW/cm².[10]

Based on information obtained from Scheme I and Figure 2, it is projected that a mixture of Trithiol and vinyl acrylate should proceed exclusive by addition of thiol to the vinyl group resulting in the formation of a highly crosslinked network. The evaluation of the polymerization mechanism of a Trithiol and vinyl acrylate equal molar mixture can be achieved by the examination of the molar consumption rates of each functional group during the polymerization in the absence of an external photoinitiator. Accordingly, Figure

3 shows relative mole conversions derived from following each functional group (acrylate, vinyl, and thiol) by real-time infrared spectroscopy as a function of polymerization time. [The samples used in plots (A) and (B) have 20 mole % and 50 mole % of Trithiol based on functional group, respectively]. According to the result in Figure 3, the acrylate group polymerizes faster than the vinyl group for both samples. Interestingly, in both cases, the thiol group conversion is identical to vinyl group conversion. The identical conversion of vinyl and thiol group, irrespective of the thiol concentration, indicates that thiol groups react exclusively with vinyl group in this system. Indeed this has been borne out by analysis of the actual products of mixtures of monofunctional thiols and vinyl acrylate.

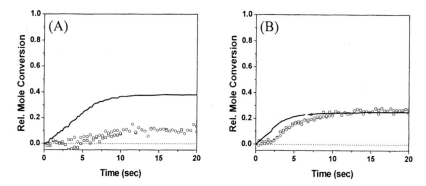

Figure 3. RTIR relative mole conversion of (—) acrylate, (O) vinyl, and (□) thiol groups versus irradiation time of vinyl acrylate/trithiol mixtures without external photoinitiators. Shown are results for (A) 20 mol% Trithiol and (B) 50 mol% Trithiol. Light intensity (full arc) is 35 mW/cm².

Evolution of the polymer structure during polymerization of the Trithiol/vinyl acrylate mixtures can be obtained by considering the polymerization behavior of the three functional groups. Based on the results in Figure 3, it is seen that even in the presence of thiol, acrylate linear polymer will be formed first as depicted in Scheme I. Then, as shown in Scheme III, a crosslinked network will be obtained by the reaction between the pendant vinyl and thiol groups causing gelation and crosslinking. As shown in Figure 3, the sample containing 50 mole % Trithiol exhibits 0.25 relative mole conversion of both the vinyl and thiol groups while the sample containing 20 mole % Trithiol has approximately 0.1 relative mole conversion indicating different crosslink density of the two samples, i. e., the sample with the higher thiol concentration undergoes a greater degree of crosslinking reaction between the thiol and pendant vinyl ether groups on the acrylate side chain.

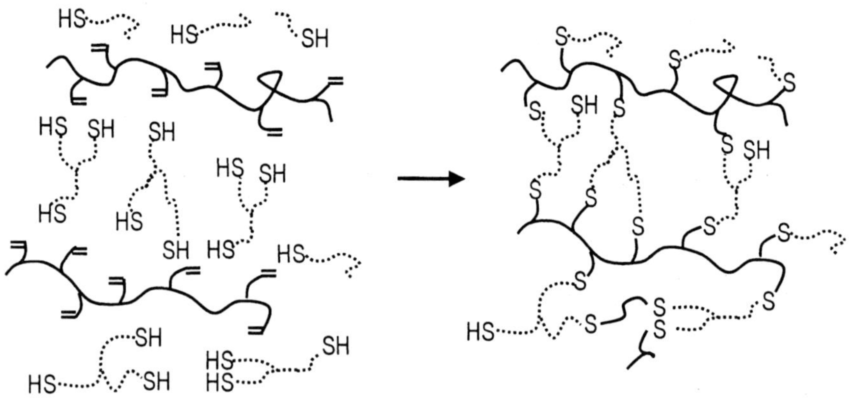

Scheme III. Schematic diagram of the evolution of the polymer network structure during copolymerization of Trithiol/vinyl acrylate mixture.

While the studies of Trithiol/vinyl acrylate copolymerization kinetics in nitrogen are valuable, polymerization of the same systems in air provides even more practical information since most photocuring processes are conducted in air. In the presence of air, most free-radical polymerizations are severely inhibited because both the initiator and growing polymeric radicals react with oxygen. The rate constant for reaction of oxygen with carbon centered radicals is much greater than initiation and propagation rate constants.[19,20] Therefore, photocuring of traditional acrylate based resins usually requires increased UV light intensity and/or increased photoinitiator concentration to overcome oxygen inhibition. However, thiol-ene polymerization exhibits significantly reduced oxygen inhibition due to an oxygen scavenging chain reaction as shown in Scheme IV. In the presence of oxygen, growing radicals are immediately scavenged by oxygen resulting in the generation of stable peroxy radicals. Subsequently the peroxy radicals abstract the hydrogen form thiols producing another thiyl radical which can reinitiate polymerization. Therefore, radical concentration is not changed and oxygen present in samples is rapidly consumed resulting in only a minor effect from oxygen inhibition.

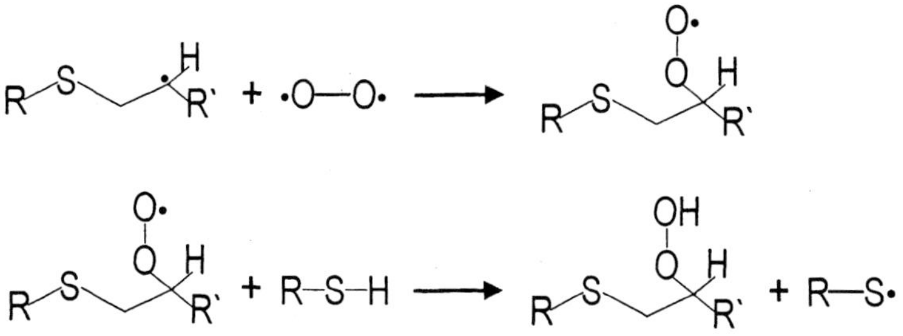

Scheme IV. Reduced oxygen inhibition mechanism by thiol

Figure 4 shows conversion versus time results for a Trithiol/vinyl acrylate mixture containing 30 mole % of Trithiol in air. Even without external photinitiators and in an air atomsphere, the mixture polymerizes quite rapidly due to the self-initiating ability of vinyl acrylate and reduced oxygen inhibition by thiol. The presence of the thiol limits oxygen inhibition as vinyl acrylate will not polymerize under the conditions employed in Figure 4 in air. Interestingly, the polymerization tendency of the 30 mole % Trithiol/vinyl acrylate mixture in air is identical to that observed in nitrogen (results in nitrogen are not shown). Again as observed for the results in Figure 3, the acrylate double bond polymerizes faster than the vinyl ene group, and thiol conversion is identical to the vinyl ene conversion.

Based on the polymerization results in Figures 3 and 4, we conclude that Trithiol/vinyl acrylate polymerization proceeds by two separate free-radical chain process, acrylate homopolymerization which forms linear acrylate polymer and thiol-vinyl group reactions which lead to the formation of a crosslinked network. The exclusive acrylate homopolymerization in the presence of the thiol is due to the very low electron density of the acrylate double bond leading to a much faster propagation rate than for traditional alkyl acrylates and a reduction in the propensity for undergoing the thiol-ene reaction with the thiol.[10] The exclusive reaction of the thiol with the vinyl ester ene is fostered by the high electron density of the vinyl double bond.

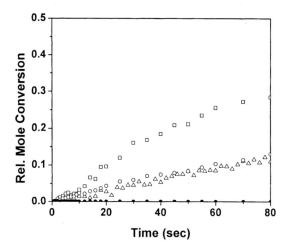

Figure 4. RTIR relative mole conversion of (—) acrylate, (○) vinyl, and (□) thiol groups versus irradiation time of vinyl acrylate/trithiol mixtures without external photoinitiators in air. Shown are results for (A) 30 mol% of Trithiol and (●) exhibit acrylate group conversion in the absence of Trithiol. Light intensity (full arc) is 35 mW/cm².

Conclusions

The polymerization of a mixture of a trifunctional thiol and vinyl acrylate in the absence of external photoinitiators has been investigated using real-time FTIR. Trithiol/vinyl acrylate mixtures readily undergo polymerization even without external photoinitiators due primarily to the ability of vinyl acrylate to generate free-radicals that intiate radical chain processes. The acrylate group of the vinyl acrylate shows a much faster polymerization rate than that of the vinyl group. Addition of a multifunctional thiol to vinyl acrylate results in enhanced vinyl group conversion by a chain transfer/free-radical addition process involving vinyl ester ene and thiol groups. Based on the mole conversion versus time results, two separate free-radical polymerization processes, an acrylate homopolymerization and a thiol-vinyl ester copolymerization, have been shown to occur simultaneously. Neither polymerization process is inhibited by oxygen.

Acknowledgements

The authors would like to acknowledge an NSF Tie Grant (0120943) in cooperation with the NSF IU/CRC Coatings and Photopolymerization Centers and Fusion UV Curing Systems for financial support.

References

1. Fouassier, J.P. *Photoinitiation Photopolymerization and Photocuring; Fundamentals and Applications*, Hanser Publishers, New York 1995.
2. Jacobine, A.T. in *Radiation Curing in Polymer Science and Technology, Volume I; Polymerization Mechanisms*; Fouassier JP, Rabek JF, eds. Elsevier Applied Science, London 1993.
3. Roffery, C.G. *Photopolymerization of Surface Coatings*, Wiley Interscience, New York 1982.
4. Pappas, S. P. *Radiation Curing, Science and Techonology*, Plenum Press, New York 1992.
5. Lowe, C.; Oldring, P.K.T. *Chemistry and Techonology of UV and EB Formulations for Coatings, Inks and Paints*, SITA Technology Ltd., London 1991.
6. Andersson, H.; Hult, A. *J. Coat. Technol.* **1997**, *69(865)*, 91.
7. Kudyakov I.V., Mattias. W. Turro, N.J. *J.Phys. Chem. A* **2002**, *106(10)*, 1938.
8. Kudyakov, I.V., Fos, W.S., Purvis MB. *Ind. Eng. Chem. Res.* **2001**, *40(14)*, 3092.
9. Lee, T.Y.; Roper, T. M.; Jonsson, E.S. ; Kudyakov, I.; Viswanathan, K.; Nason, C.; Guymon, C.A.; Hoyle, C. E. *Polymer* **2003**, *44(10)*, 2859.
10. Lee, T. Y.; Roper, T. M.; Jonsson, E. S.; Guymon, C. A.; Hoyle, C. E. *Macromolecules* **2004**, *37(10)*, 3606.
11. Morgan, C. R.; Ketley, A. D. *J. Polym. Sci. Polym. Lett. Ed.* **1978**, *16*, 75.
12. Gush, D. P.; Ketley, A. D. *Modern Paint and Coatings* **1978**, *11*, 68.
13. Fouassier, J.P.; Rabek, J. F. *Radiation Curing in Polymer Science and Technology-Volume I; Polymerization Mechanism*, Elsevier Applied Science, London, 1993.
14. Hoyle, C. E.; Lee, T. Y.; Roper, T. M. *J. Polym Sci Part A: Polym Chem.* **2004**, *42*, 5301.
15. Cramer, N. B.; Scott, J. P.; Bowman, C. N. *Macromolecules,* **2002**, *35*, 5361.
16. Cramer, N. B.; Bowman, C. N. *J. Polym. Sci., Part A: Polym. Chem.* **2001**, *39*, 3311.

17. Cramer, N. B.; Davies, T.; O'Brien, A. K.; Bowman, C. N. *Macromolecules* **2003**, *36*, 4631.
18. Cramer, N. B.; Reddy, S. K.; O'Brien, A. K.; Bowman, C. N. *Macromolecules* **2003**, *36*, 7964.
19. Turro, N. J. *Modern Molecular Chemistry*, Univ. Sci. Books, Mill Valley, CA 1991.
20. Odian G. *Principles of Polymerization*, 3rd ed., Wiley Interscience, New York 1982.

Chapter 3

Photopolymerization Kinetics of Pigmented Systems Using a Thin-Film Calorimeter

Todd M. Roper[1], C. Allan Guymon[2], and Charles E. Hoyle[1,*]

[1]Department of Polymer Science, University of Southern Mississippi, Southern Station Box 10076, Hattiesburg, MS 39406
[2]Department of Chemical and Biochemical Engineering, University of Iowa, Iowa City, IA 52242

A thin-film calorimeter was used to evaluate the polymerization rates of pigmented photopolymerizable systems. As a result of a wide linear response range, which is more than an order of magnitude greater than that of the photo-DSC, the thin-film calorimeter is capable of measuring both small and large signals accurately. It was used to measure the polymerization exotherms of thin-films of a photocurable acrylate monomer with added pigment as well as commercial UV curable pigmented ink formulations.

Introduction

Photopolymerization is a field experiencing rapid growth due to its versatility, low environmental impact, and the production of crosslinked films and coatings with excellent material performance properties.[1,2] The discovery of new applications, including thin-films and pigmented inks,[3,4] has been driven by recent research identifying the structural characteristics responsible for increased rates of reaction[5-8] and reduced oxygen inhibition.[9-12] Unfortunately, progress in these emerging applications has been hindered by the absence of analytical instrumentation capable of quantitatively measuring the photopolymerization kinetics. The construction of versatile tools which are efficient, simple to use, inexpensive, and capable of accurately evaluating the polymerization kinetics of thin-films and pigmented systems would significantly promote future technological advances.

Key factors affecting the film formation process for a photopolymerizable coating include the polymerization mechanism and kinetics. Measurement of polymerization kinetics has been achieved in the past using a variety of different methods including real-time spectroscopy (IR and Raman),[7,13-15] rheology,[16-18] pyrommetry,[19-22] and calorimetry.[23-27] Photocalorimetry, which involves the measurement of heat evolved during a chemical reaction initiated by a photon of light, affords a direct sensitive measurement of the polymerization rate. The polymerization rate is calculated by assuming that the amount of heat released is directly proportional to the number monomer units converted to polymer. Photo differential scanning calorimetry (photo-DSC) is one of the most commonly used calorimetric methods for examining bulk polymerization kinetics of unfilled systems.[23-27] Despite its widespread use, photo-DSC is plagued by poor time resolution (on the order of seconds), the requirement of large sample volumes (not applicable to thin films), and limitation to the analysis of nonvolatile monomers. In addition, accuracy and reproducibility are problems because the exotherm curve is dependent on the distribution of sample in an aluminum sample pan. Also, cost prevents its usage in many industrial settings.

A photocalorimeter based upon a single thin-film heat flux sensor was first developed by Wisnosky and Fantazier.[28,29] Additional use of the thin-film calorimeter with minor modifications of the original modification has since been reported.[30-32] A new higher performance version of the thin-film calorimeter (TFC) with significantly improved operational characteristics has been described in detail in a very recent publication.[33] The major improvements to Wisnosky and Fantazier's original thin-film calorimeter include significantly enhanced sensor sensitivity, increased data acquisition resolution, and considerably greater sampling rates.[33] As a result, the TFC has the ability to accurately measure a much smaller stimulus, hence making the quantitative analysis of photopolymerizable systems which evolve a small amount of heat possible. The superior time response and increased sensitivity of the thin-film sensor suggests many other applications. Herein, we report the use of a TFC for measuring the photopolymerization kinetics of pigmented thin-films and pigmented coatings.

Experimental

Instrumentation

A photo differential scanning calorimeter (photo-DSC), used to collect the photopolymerization exotherms for comparison with the TFC, was constructed from a Perkin Elmer DSC 7 by modifying the head with quartz windows. A 450 Watt medium-pressure mercury lamp from Ace Glass was shielded from the DSC head by an electric shutter. Samples of 2.00 ± 0.03 mg were added to crimped aluminum sample pans and purged with nitrogen for four minutes prior to irradiation. Sample thicknesses were ~175 to 200 μm.

A Bruker IFS 88 modified to accommodate a horizontal sample accessory was used to collect real-time infrared (RTIR) spectroscopic data to confirm/compliment kinetic results collected using the TFC. The experimental set-up has been previously described in detail.[3,7] The acrylate IR absorbance band used to monitor the disappearance of reactant/monomer during the photopolymerization was 812 cm⁻¹. A Cary 500 Scan UV-Vis spectrometer was used to collect pigment absorbance spectra. The pigment was dispersed in acetonitrile solvent on a ball mill using ceramic grinding aids.

Thin-Film Calorimeter

A schematic of the thin-film calorimeter (TFC) is shown in Figure 1. A flexible thin-film heat flux sensor from Omega Engineering is attached via

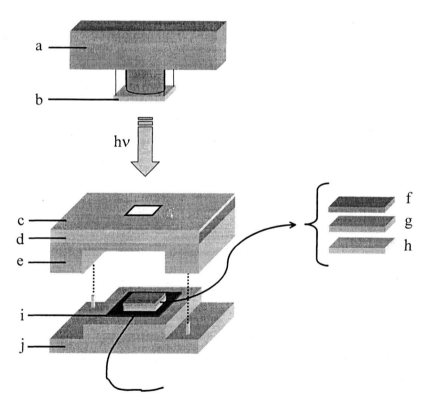

Figure 1. A partial diagram of the thin-film calorimeter (TFC) including the (a) lamp system, (b) electric shutter, (c) quartz heat shield, (d) insulating cork, (e) machined aluminum cap, (f) insulating cover slip, (g) sample, (h) protective substrate, (i) thin-film heat flux sensor, and (j) machined aluminum heat sink.

double sided adhesive tape to a metal heat sink. Adjustable clamps hold a lid containing a 1 cm^2 hole onto the heat sink. A 450-Watt medium pressure mercury arc lamp with an electric shutter is positioned above the sensor. The sensor is connected to a battery powered DC micro-voltmeter. The signal is filtered, amplified, and converted to a digital signal by an A/D converter board interfaced with a PC.

A protective barrier such as aluminum (Al) or polyethylene (PE) must be used between the sample and sensor to prevent damage to the sensor. Thin pigmented films were prepared by solvent casting the inks onto the Al/PE support. Acetone was used as the solvent and the ultimate film thickness depended on the concentration of solvent employed during the casting process (5-20 micron). The coated support was placed on the sensor and the metal lid clamped onto the heat sink.

The sensor consists of 40 thermopiles or measurement sites, which are connected in series and encased in a polyimide film. No external voltage or current stimulation is required to operate the sensor, and the measured heat flux (voltage output) is readable by a voltmeter with microvolt resolution. The voltmeter is interfaced with a PC through a multichannel A/D converter having 16 bit resolution and a maximum sampling rate of 100 kHz. As constructed, the TFC has the capacity to collect 100,000 voltage measurements per second; however, the data are usually averaged to 100 data points per second to enhance the signal-to-noise ratio. A complete description of all aspects of the calibration and design of the thin-film calorimeter is given in reference 33.

Materials

The materials used in this study include the monomers 1,6-hexanediol diacrylate (HDDA) and tripropylene glycol diacrylate (UCB Chemicals Corp.), and the photoinitiator 2,2-dimethoxy-2-phenyl acetophenone (Ciba Specialty Chemicals). A red organic pigment, calcium lithol rubine, and assorted UV curable ink formulations were also used (Flint Ink Corporation). All chemicals were used as received, without additional purification. The respective chemical structures are shown in Figure 2.

(a)   (b)   (c)

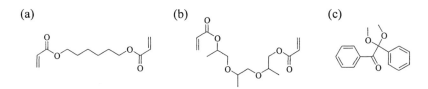

Figure 2. Chemical structures of the acrylate monomers and photoinitiator used in test formulations; (a) 1,6-hexanediol diacrylate (HDDA), (b) tripropylene glycol diacrylate(TRPGDA), (c) 2,2-dimethoxy-2-phenyl acetophenone

Results and Discussion

As already described in the Experimental section, the TFC records polymerization exotherms by measuring the heat released as a function of time. Valuable information regarding the polymerization process, such as rate and extent of conversion, can be obtained from the polymerization exotherm. As an example, Figure 3 shows the photopolymerization exotherm from a 100 μm thick HDDA sample with 1 wt% DMPA photoinitiator. Upon irradiation, there is a brief inhibition period due to the presence of dissolved oxygen or impurities inhibiting the polymerization. The polymerization exotherm values are proportional to the polymerization rate at each point during the reaction, with the exotherm maximum being proportional to the maximum polymerization rate.

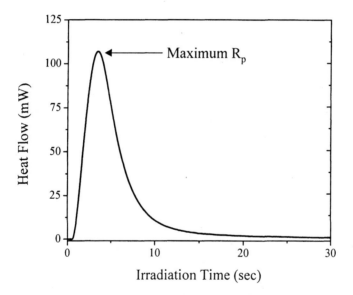

Figure 3. TFC photopolymerization exotherm for a 25 micron thick HDDA film containing 1 wt% DMPA. The light intensity was 20 mW/cm².

An instruments' sensitivity is a critical performance feature since it imposes limitations regarding the types of systems it can accurately examine. To illustrate the excellent sensitivity of the TFC, a comparison is made to the traditionally photo-DSC. Free-radical photopolymerization kinetic equations dictate that the polymerization rate is proportional to light intensity to the 0.6-0.7 power.[30] Therefore, the linear range in a log-log plot of the polymerization rate as a function of light intensity approximates the range of heat flux that analytical tools can accurately measure. Figure 4 clearly shows that linearity is maintained with the TFC over a light intensity range much larger than the photo-DSC when samples of HDDA with 1 wt% DMPA are polymerized at various light intensities. Remarkably, the TFC is able to measure polymerization exotherms conducted at both lower and higher light intensities, which corresponds to the measurement of smaller and larger quantities of heat flux. Small quantities of heat are released in systems with low concentrations of reactive groups, as is the case in many commercial applications of UV curing such as thin, clear films and pigmented inks. Large quantities of heat are released by extremely rapid polymerizing monomers and/or thick coatings.

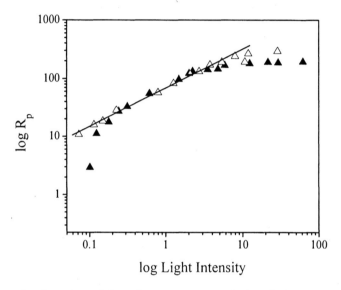

Figure 4. *log-log plots of the polymerization rate (R_p) vs. light intensity for 200 micron thick HDDA samples using the (▲) photo-DSC and (△) TFC. All samples contain 1 wt% DMPA.*

Photopolymerizable inks represent a class of applications which are characterized by the presence of pigments. The pigments provide color by selectively absorbing/scattering incident visible light (380-780 nm). In addition, pigments also absorb/scatter incident ultraviolet radiation (180-380 nm), which results in a reduction of the polymerization rate from the competition for available photons between the pigment and photoinitiator. The extinction coefficient of the pigment in the ultraviolet region of the electromagnetic spectrum depends on the wavelength and crystalline structure of the pigment. Figure 5A shows the UV-Vis absorbance spectra for organic blue, yellow, and red pigments. The red pigment was used in this work because it had the lowest absorbance at 365 nm, which is one of the major emission bands of a mercury arc lamp used as the excitation source during polymerization. Figure 5B shows the UV-Vis absorbance spectra of the red organic pigment at various concentrations. An increase in the pigment concentration will result in an increase in the overall brightness/intensity of the color and hiding (increase in absorbance of visible wavelengths) at the expense of the polymerization rate (increase in absorbance of ultraviolet wavelengths). As a result, photocurable inks are generally applied as thin-films in order to obtain the desired optical properties while maintaining an adequate polymerization rate and achieving through cure.

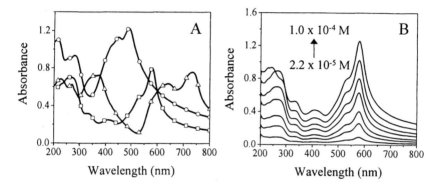

Figure 5. UV-Vis absorbance spectra of (A) a organic blue (△), yellow (○), and red (□) pigment at a concentration of 0.00007M in acetonitrile and (B) the red pigment at an increasing concentration in acetonitrile.

Calorimetric analysis of thin-film photopolymeriz ations is impossible to achieve using traditional equipment. Aside from its poor sensitivity, it is not easy to measure the polymerization exotherms of thin-films with the photo-DSC because of the difficulty in uniformly covering a DSC pan: extremely thick samples approaching 175 micrometers are generally used. The inability to accurately examine thin-film photopolymerization kinetics makes analysis of photopolymerizable inks very difficult. To illustrate this point, polymerization exotherms of pigmented systems were collected on both the photo-DSC and TFC at a thickness between 175-200 microns as shown in Figure 6. At a loading of only 5 wt% pigment (photopolymerizable inks may have up to 20 wt% pigment or more), the polymerization exotherm cannot be measured on the photo-DSC because the rate is extremely slow (thick sample) and the instrument does not have the requisite sensitivity: see previous discussion of this point. On the other hand, the TFC is able to record the polymerization exotherm:

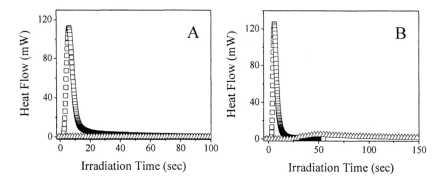

Figure 6. Photopolymerization exotherms collected on the (A) Photo-DSC and (B) TFC for a TRPGDA sample with (△) and without (□) 5 wt% of red pigment. All samples contain 1 wt% of the photoinitiator, DMPA, and exposed to constant irradiation having an intensity of 10 mW/cm².

interestingly, due to the extreme thickness of the pigmented sample, polymerization does not occur throughout the entire film and only a low exotherm is recorded.

In addition to its utility in evaluating thick pigmented systems, which in reality are not used commercially, the TFC is especially useful for evaluating the photopolymerization characteristics of thin-films/pigmented systems because of its' inherent sensitivity. Figure 7 depicts the raw photopolymerization exotherm result for a pigmented system (TRPGDA with 15 wt% red pigment) at a thickness of 10 microns. Since the TFC has no reference cell, the background heat from the light when dealing with thin samples is substantial and must be subtracted from the raw photopolymerization exotherm. Figure 7 shows the experimental sequence which must occur when collecting a photopolymerization exotherm of a thin pigmented system using the TFC. In step (I), the thin-film is constantly irradiated with high energy light such that polymerization occurs. After the polymerization has ceased, the light source is removed in step (II) until the signal has reached equilibrium. Finally, the polymerized film is constantly irradiated with light having the same intensity as step (I) in order to measure the background heat. As shown in Figure As shown in Figure 8A, the heat from polymerization can be easily seen when superimposing the steps (I) & (III) from Figure 7. By simply subtracting the background heat from the convoluted polymerization exotherm, the real photopolymerization exotherm for a pigmented system is obtained as shown in Figure 8B.

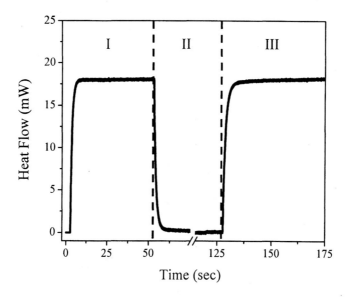

Figure 7. A TFC photopolymerization sequence consisting of the (I) photopolymerization exotherm, (II) dark period, and (III) background heat. Photopolymerization exotherm of 10 μ TRPGDA sample with 15 wt% red pigment and 1 wt% DMPA irradiated with 10 mW/cm² light.

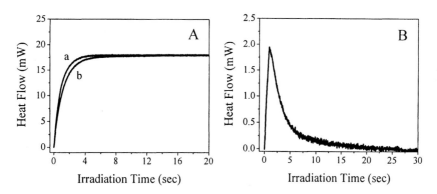

Figure 8. (A) Actual recorded photopolymerization exotherm (a) and the background heat (b) for a 10 micron thick TRPGDA film containing 15 wt% red pigment and 1 wt% DMPA upon irradiation with light (full arc) having an intensity of 10 mW/cm². (B) Extracted photopolymerization exotherm.

38

Commercial photopolymerizable inks consist of pigment dispersed in reactive oligomers. The concentration of reactive groups is lower for oligomeric reactants than monomeric reactants and, as a consequence, small quantities of heat are released upon polymerization making characterization difficult. Figures 9A and 10A show the TFC polymerization exotherm of a series of yellow and black photopolymerizable ink formulations provided by Flint Ink Corporation, respectively. As shown in Figure 9A, the polymerization exotherms of the yellow ink samples (in curves a and b) were much larger than a third sample (curve c), indicating faster polymerizations. The corresponding RTIR results shown in Figure 9B compliment and confirm the TFC based conclusions. Similarly, the RTIR results in Figure 10B confirm the observed TFC trends in Figure 10A for three black photocurable inks samples. These results indicate that due to the high sensitivity and flexibility in the experimental set-up (film thickness control), the TFC has the potential to aid in future developments in the field of photopolymerization.

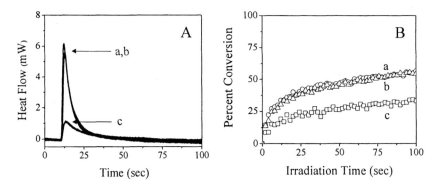

Figure 9. Polymerization exotherms of three yellow photocurable inks samples recorded on (A) the TFC (curves a, b, and c) , and (B) the RTIR (corresponding curves a, b, and c). The TFC results were obtained on 8 micron thick samples using light from a medium pressure mercury lamp having and intensity of 40 mW/cm². The RTIR results were obtained on 20 micron thick samples using light from a high pressure mercury lamp having a light intensity of 40 mW/cm².

Conclusions

A thin-film calorimeter was used to record the polymerization exotherms of pigmented photocurable mixtures. The exotherms of both thick and thin pigmented films were successfully recorded. The results for the thin-film calorimeter were corroborated by real-time infrared measurements.

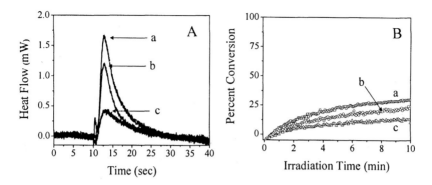

Figure 10. Polymerization exotherms of three black photocurable inks samples recorded on (A) the TFC (curves a, b, and c), and (B) the RTIR (corresponding curves a, b, and c). The TFC results were obtained on 8 micron thick samples using light from a medium pressure mercury lamp having and intensity of 40 mW/cm². The RTIR results were obtained on 20 micron thick samples using light from a high pressure mercury lamp having a light intensity of 40 mW/cm².

Acknowledgements

The authors would like to acknowledge the NSF IU/CRC Coatings Center, the NSF Photopolymerization Center, and an NSF Tie Grant (0120943) for financial support.

References

1. Fouassier, J. P. Photoinitiation Photopolymerization and Photocuring; Fundamentals and Applications; Hanser Publishers: Munich, 1995.
2. Pappas, S. P. *Radiation Curing Science and Technology*; Plenum Press: New York, 1992.
3. Roper, T. M.; Kwee, T.; Lee, T. Y.; Guymon, C. A.; Hoyle, C. E. *Polymer* **2004**, *45*, 2921.
4. Roper, T. M.; Rhudy, K. L.; Chandler, C. M.; Guymon, C. A.; Hoyle, C. E. Radtech US 2002 Technical Proceedings, Indianapolis, IN; p. 697.
5. Decker, C.; Moussa, K. *Makromolecular Chem* **1991**, *192*, 507.
6. Decker, C.; Moussa, K. *European Polymer Journal* **1991**, *27*, 881.

7. Lee, T. Y.; Roper, T. M.; Jonsson, S. E.; Kudyakov, I.; Viswanathan, K.; Nason, C.; Guymon, C. A.; Hoyle, C. E. *Polymer* **2003**, *44(10)*, 2.
8. Lee, T. Y.; Roper, T. M.; Jonsson, S. E.; Guymon, C. A.; Hoyle, C. E. *Macromolecules* **2003**, *37*, 3659.
9. Jacobine, A.F. In Radiation Curing in Polymer Science and Technology III, Polymerization Mechanisms; Fouassier, J.D.; Rabek, J.F., Eds.; Elsevier Applied Science: London, 1993; Chapter 7, p. 219.
10. Morgan, C. R.; Ketley, A. D. *J Rad Cur* **1980**, *7(2)*, 10.
11. Morgan, C. R.; Magnotta, F.; Ketley, A. D. *J Polym Sci* **1977**, *15*, 627.
12. Lee, T.Y.; Guymon, C. A.; Jonsson, E. S.; Hoyle, C. E. *Polymer* **2004**, *45*, 6155.
13. Decker, C.; Moussa, K. *Macromolecules* **1989**, *22*, 4455.
14. Anseth, K.S.; Decker, C.;, Bowman, C.N. *Macromolecules* **1995**, *28*, 4040.
15. Cramer, N.B.; Bowman, C.N. *J Polym Sci Part A: Polym Chem* **2001**, *39*, 3311.
16. Chiou, B.; English, R.J.; Khan, S.A. *Macromolecules* **1996**, *29*, 5368.
17. Chiou, B.; Khan, S.A. *Macromolecules* **1997**, *30*, 7322.
18. Chiou, B.; Raghavan, S.R.; Khan, S.A. *Macromolecules* **2001**, *34*, 4526.
19. Falk, B.; Santigo, M.; Crivello, J.M. *J Polym Sci Part A: Polym Chem* **2003**, *41*, 579.
20. Crivello, J.V.; Falk, B.; Zonca, M. R. Jr.; *J Applied Polym Sci* **2004**, *92(5)*, 3303.
21. Falk, B.; Vallinas, S. M.; Crivello, J. V. *Polym Mat Sci Eng* **2003**, *89*, 279.
22. Falk, B.; Vallinas, S. M.; Crivello, J. V. *Polym Mat Sci Eng* **2003**, *88*, 209.
23. Tryson, G.R.; Schultz, A.R. *J Polym Sci: Polym Physics Ed* **1979**, *17*, 2059.
24. Boots, H.M.J.; Kloosterboer, J.G.; Van Die Hei, G.M.M. *British Polym J* **1985**, *17(2)*, 219.
25. Hoyle, C.E. In Radiation Curing Science and Technology; Pappas, S.P., Plenum Press: New York, 1992; pp. 57-130.
26. Lester, C.; Guymon, C.A. *Polym* **2002**, *43(13)*, 3707.
27. McCormick, D. T.; Stovall, K. D.; Guymon, C. A. *Macromolecules* **2003**, *36(17)*, 6549.
28. Wisnosky, J.D.; Fantazier, R.M. *Journal of Radiation Curing* **1981**, *8(4)*, 16.
29. Fantazier, R.M. U.S. Patent 4,171,252 **1979**.
30. Hoyle, C.E.; Hensel, R.D.; Grubb, M.B. *Polymer Photochemistry* **1984**, *4*, 69.
31. Pargellis, A.N. *Rev. Sci. Instrum.* **1986**, *57(7)*, 1384.
32. Hager Jr., N.E. *Rev Sci Instrum* **1987**, *58(1)*, 86.
33. Roper, Todd: Hoyle, Charles E.: Guymon, C. Allan *Rev Sci. Inst.* **2005**, *76(5)*, 1.

Chapter 4

Desorption of Surfactants During Film Formation

F. Belaroui[1], B. Cabane[2], Y. Grohens[3], P. Marie[1], and Y. Holl[1,*]

[1]Institut Charles Sadron (CNRS) and Université Louis Pasteur, B.P. 40016,
67083 Strasbourg, France
[2]Ecole Supérieure de Physique et Chimie Industrielle, 10, Rue Vauquelin,
75005 Paris, France
[3]Université de Bretagne Sud Laboratoire Polymères, Propriétés aux
Interfaces and Composities, Rue de Saint-Maudé, 56 321 Lorient, France

Desorption of surfactants during latex film formation is due to confinement when particles get very close upon colloid drying. In a first part, evidence for desorption is briefly presented through Raman Confocal Spectroscopy and optical microscopy results showing surface enrichment or segregation in separated domains. A second part is devoted to specific studies of the surfactant desorption event. For that purpose, four monodisperse core-shell latices were synthesized and deuterated SDS together with special contrast conditions were used in Small Angle Neutron Scattering experiments. It was shown that desorption occurred early in the process, when the latex still contained around 20 % of water. A small fraction of the surfactant remained irreversibly adsorbed at the particle surface.

Introduction

Latex films very often also contain small molecules, intentionally or not. These small molecules can be, for instance, undesirable residual monomers or solvent; or they can be voluntary added formulation additives like stabilizers, plasticizers, coalescing aids, adhesion promoters, crosslinking agents and catalysts. In all cases, these molecules have a profound influence on properties. Not only do the small molecules affect properties through their total amount, but also, at a given concentration, through their distribution in the material. Distribution can be homogeneous or heterogeneous with enrichment at the film – air or film-substrate interfaces or accumulation into aggregates in the bulk of the film. If a small sized molecule like a surfactant accumulates at the film-substrate interface of a waterborne coating, adhesion properties, especially in wet conditions, are strongly affected. Other example, in food packaging (or food can varnishes) non polymeric molecules should absolutely not migrate to the surface in contact with food. As far as surfactants are concerned, it is well known that their distribution is closely related to drying mechanisms (1). However, we are far from a complete understanding of the whole process leading to a distribution and from being able to predict a distribution for a given system dried in given conditions. It appears that a key issue in this problem is surfactant desorption when particles come in close vicinity upon drying. The questions of interest are then: when does surfactant desorption occur and to what extent (depending on drying conditions, a certain amount of surfactant may remain trapped at the interface)? This paper presents first attempts to address these issues.

Materials and Methods

Latices

Four core-shell latices were synthesized, differing by the composition and Tg of the core (either pure poly(n-butyl acrylate) of very low Tg or poly(n-butyl acrylate-co-methyl methacrylate) with a Tg slightly below room temperature) and the percentage of acrylic acid in the composition (either 1 or 4 wt %). The pH directly after synthesis was 2, the solids content was 25 wt %. More details on the syntheses can be found in reference 2.

Latices were purified by dialysis using a Millipore membrane until the conductivity of the water in contact with the latex was less than 3 μS/cm.

Purification eliminates water soluble residual salts, oligomers, and the surfactant (except for traces, depending on the nature of the surfactant/polymer system (*3*)) used in the synthesis. The latices remained stable after purification because of the high concentration of acrylic acid in the shell (see below). After purification, the latices could be post-stabilized by addition of controlled amounts of SDS, labeled or not.

Directly following purification, latices were characterized by measurements of solids content (gravimetry), pH (pH could then be increased by addition of sodium hydroxide), mean particle size and particle size distribution (dynamic light scattering), differential scanning calorimetry and by determination of the acrylic acid repartition in the particles (*4*). They are fairly monodisperse, having all very similar mean diameters (close to 100 nm). Furthermore, the acrylic acid is well incorporated into the particles (only low amounts are lost in the serum) and a large majority of it is located in the particle shell (as indicated by titration with sodium hydroxyde). In this paper, representative results will be shown, mainly concerning the latex with a poly(butyl acrylate) core and 1% of acrylic acid (BuA 1). For more details about the effects of core Tg and acrylic acid content, see reference 13.

Films were formed by casting the latices onto polyethylene substrates (Parafilm) (for Raman spectroscopy) or quartz plates (for SANS measurements) and drying at 23 °C and 50 % relative humidity. The thickness of the dry films was (100 ± 5) μm.

Other acrylic latices (poly(2-ethyl hexyl methacrylate)) were also used for microscopy and DSC experiments (*5*).

Confocal Raman Spectroscopy

This technique allows one to record a Raman spectrum of a small volume (in the micrometer range) of the film. Thanks to the confocal optics, this sampling volume can be moved along the z direction from the surface (film-air interface) to the inside of the film over thicknesses of several tens of micrometers. Furthermore, the whole sample can be moved along the x and y directions allowing a complete (although tedious) mapping of the film.

Most experimental details were given elsewhere (*6*), only the main points will be recalled here. Raman measurements were performed with a Jobin Yvon, Labram confocal Raman spectrometer. Excitation at a wavelength of 632.8 nm was provided by a He-Ne laser. The power delivered to the sample was about 10 mW. Depth profiling was possible by tuning the plane of focus of the microscope step by step. Confocal Raman Spectroscopy can be used for quantitative analysis since the Raman signal is known to be proportional to the concentration. The procedure used for the determination of the concentration of small molecules in the dry latex films was described in reference 6. Error bars (not shown in Fig. 1) were estimated to 5%.

Small-Angle Neutron Scattering

Neutron scattering experiments were carried out with D11 and D22 diffractometers at the Institute Laue Langevin (ILL) at Grenoble, France. The wavelengths were $\lambda = 6$ Å or 10 Å, and two sample-to-detector distances of 35.7 m or 18 m were used, ensuring a large q range as 1.54×10^{-3} Å^{-1} < q < 3.4×10^{-2}Å^{-1}. The data were treated according to standard ILL procedures for small angle isotropic scattering (7).

Two kinds of latices were prepared by a dilution-reconcentration procedure, both at pH 10, and with a solids content of 25 wt %. The first ones contained 20 % of D_2O and 1 or 6 wt % of deuterated SDS (from Euriso-Top CEA, Saint Aubin, France), noted SDS-D (the whole hydrophobic moiety was deuterated). This concentration of D_2O in the continuous phase, determined experimentally, ensured the "index matching" between the serum and the particles. Without SDS-D, such latices did not scatter neutrons because there was no contrast between serum and particles. With SDS-D, the only source of contrast being the surfactant, neutron scattering spectra revealed the repartition of the surfactant in the drying film.

The second sort of latices contained almost 100 % of D_2O and 1 or 6 wt % of hydrogenated (ordinary) SDS, noted SDS-H. In this case, neutron scattering spectra revealed the structure of the residual water in the drying film.

These latices were dried on quartz plates at 23°C and 50 % RH. When concentrated enough, they were placed in the instrument and scattering spectra were recorded versus time, at 45, 60 and 90 min. In some cases, the latices were also dried in an oven at 75°C for 90 min.

Results and Discussion

Evidence for surfactant desorption

If the surfactant remained adsorbed at the particle surface, taking into account the sizes of the particles (around 100 nm) and of the analytical spot in Raman spectroscopy (typically 2 µm), this latter would contain several thousands of particles and the distribution of a surfactant in a film studied with this technique would appear homogeneous. The fact that it is not, that the distribution shows strong heterogeneities with surface enrichments and aggregates within the bulk is thus a first evidence for desorption. As an illustration, Figure 1 presents a concentration profile of SDS in a poly(butyl acrylate-co-acrylic acid) film established using Raman confocal spectroscopy. Both interfaces, film-air and film-substrate, are enriched with surfactant and aggregates are detected in the bulk through large peaks in the profile. This profile was taken along a vertical line (z axis) located near to the center of the film. If another line was chosen, the profile would be qualitatively similar (surface enrichment, presence of aggregates) but different mainly in the sense that aggregates would be located differently.

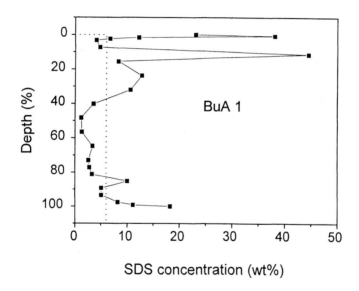

Figure 1. Confocal Raman spectroscopy data. Concentration profile of sodium dodecyl sulfate (6 wt% on average) in poly(butyl acrylate-co-acrylic acid). Drying conditions: 10 days at 23°C, 50% RH.

Using Raman confocal spectroscopy, it is also possible to map an aggregate in an x-y plane underneath the surface (Figure 2). It can be seen that quite large aggregates can be formed after surfactant desorption.

Another evidence for aggregation following desorption is provided by simple examination of films using polarized light optical microscopy (Figure 3). White spots correspond to crystallized cationic surfactant domains in poly(2-ethyl hexyl methacrylate) latex films. It is also worth noting that drying conditions influence surfactant distribution. In figure 3, faster drying (left) gives rise to larger and more numerous crystallized aggregates than slower drying (right). Slower drying allows better phase separation between surfactant and polymer. The surfactant is thus mostly expelled towards the film surface and interface. Crystallization of the segregated surfactant can also be evidenced by Differential Scanning Calorimetry (5).

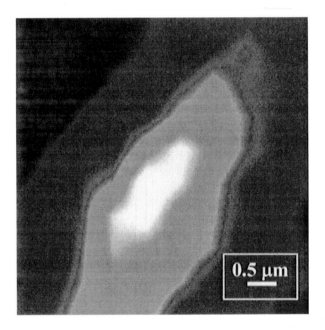

Figure 2. Confocal Raman spectroscopy data. Shape of a SDS aggregate located 15 μm under the surface in the same poly(butyl acrylate-co-acrylic acid) film as in Figure 1. Grey levels are proportional to SDS concentrations. The white domain in the center of the image corresponds to pure SDS.

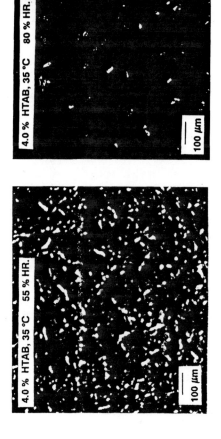

Figure 3. Polarized light optical micrographs of hexadecyl trimethyl ammonium bromide (HTAB) in poly(2-ethyl hexyl methacrylate) latex films. Fast drying on the left, slow drying on the right. White spots are crystallized surfactant domains.

Specific studies of surfactant desorption using SANS

Figure 4 shows the results of SDS desorption studies. The film containing 1 wt % of SDS-D (Figure 4, top) exhibits a strong scattering peak at a q value corresponding to the particle diameter after 45 min of drying at 23°C (solids content: 80 %). This means that the surfactant is still at the particle surface at that time. At 60 min (solids content: 85 %), the peak has almost disappeared, it is only faintly detectable. In this system, most of the surfactant leaves the particle surface at solids contents between 80 and 85%; however, small amounts of SDS remain at the surface, also after 90 min. Desorption of the major part of the surfactant occurs in a narrow time (or solids content) slot. No peak is visible after annealing at 75°C indicating that this treatment leads to a total desorption of the surfactant.

The spectra of the film containing SDS-H (Figure 4, bottom) show scattering peaks after 45 to 90 min of drying at 23°C, indicating the presence of thin water layers between particles. Annealing at 75°C has the consequence of suppressing the scattering peak at the q value corresponding to the particle diameter. Following this treatment, total expulsion of water from the interfaces has occurred.

Results with 6 wt % of SDS are qualitatively identical: traces of SDS remain trapped at the interface after 90 min of drying and are only displaced by annealing whereas total coalescence has not yet occurred after 90 min at 23°C and takes place after annealing at 75°C.

Surfactant desorption from latex particles is often studied in the wet stage to follow competition for the particle-water interface between surfactants or between surfactants and polymers (8, 9) or to check for the efficiency of latex purification procedures (10). Results are tentatively rationalized in terms of free energy of adsorption (8). A fraction of surfactant remaining irreversibly adsorbed is often observed as, for example, the non ionic Triton X-100 on polystyrene particles (11).

Desorption during film formation, in highly concentrated latices, is a different issue although sharing common features with the case referred to in the preceding paragraph. It is only scarcely studied. Joanicot et al. (12) have discussed the problem in the framework of coalescence and stressed the influence of the core mobility. Kientz et al. (5) published an article specifically devoted to surfactant desorption during film formation. They evidenced desorption through surface enrichment, surfactant segregation in separated domains followed by crystallization, surfactant concentration increase in the water phase in the special case of film formation in an excess of water. They also showed an unexpected situation where the surfactant did not desorb and increased the rigidity of the interparticular zone throughout the film. The mechanical properties of the film were dramatically affected.

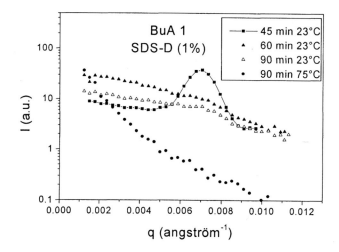

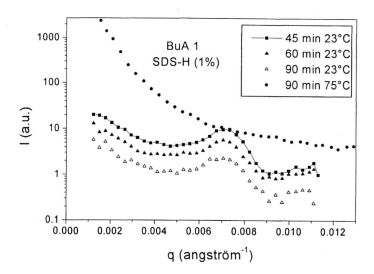

Figure 4. SDS desorption studies by SANS. Scattered intensity versus q for films containing 1 wt % of SDS-D (top) or SDS-H (bottom). In the SDS-D latex there was contrast matching between the polymer and the serum. In the SDS-H latex, D_2O was used. Vertical positions of the spectra have no significance. Drying conditions as indicated.

By using labeled SDS, the present study shows that the surfactant is desorbed very early in the film formation process, before total drying and far before breaking of the acrylic acid rich shell (Fig. 4). Surfactant desorption is most probably caused by repulsion between ionic heads facing each other when the adsorbed surfactant layers get very close. One can think of several parameters controlling the desorption event: adsorption energy, interface curvature, repulsive pressure between surfaces, shear by water flux in the tiny interparticular spaces,... Obviously, this research deserves to be carried on in order to determine the hierarchy of these parameters.

Traces of SDS remain at the interface and are probably trapped inside the acrylic acid rich domains upon coalescence (13). Residual water remaining present at the interfaces between particles after most of the surfactant has desorbed (Fig. 4, bottom) corresponds mainly to hydration of acrylic acid moieties of the shell polymer and sulfate groups from the trapped surfactant heads.

The above results pertain to SDS. Different surfactants are likely to behave differently, especially non ionic ones where the repulsive interactions between heads are weaker and shorter ranged.

Conclusion

At this point, it is clear that more systematic data are required in order to investigate several parameters of interest, namely the nature of the polymer and surfactant, the surfactant concentration, the serum characteristics (pH and ionic strength) and the drying rate. SANS is a powerful technique but not well suited for systematic data acquisition. Currently, we are developing two other methods: fluorescence decay and surface force (SFA) measurements.

References

1. Guigner, D.; Fischer, C.; Holl, Y. *Langmuir* **2001**, 17, 6419.
2. Belaroui, F. *PhD thesis, University of Mulhouse*, France, 2002.
3. Weng, L.T.; Bertrand, P.; Stone-Masui, J.H.; Stone, W.E.E.; Briggs, D.; Pignataro, S. *Surf. Interface Anal.* **1994**, 21, 387.
4. El-Aasser, M.S.; Loncar Jr, F.V.; Vanderhoff, J.W. in "Emulsion Copolymerization", (J. Guillot and C. Pichot Eds.), p. 335, Huthig and Wepf Verlag: Basel, Heidelberg, New York, 1985.
5. Kientz, E.; Dobler, F.; Holl, Y; *Polym. Int.* **1994**, 34, 125.

6. Belaroui, F.; Grohens, Y.; Boyer, H.; Holl, Y. *Polymer* **2000**, 41, 7641.

7. Ghosh, R. E.; Egelhaaf, S. U.; Rennie, A. R. *A Computing Guide for Small-Angle Scattering Experiments, ILL Internal Report* 98GH14T, 1998.

8. Colombie, D.; Landfester, K.; Sudol, E.D.; El-Aasser, M.S.; *Langmuir* **2000**, 16, 7905.

9. Kling, J.A.; Ploehn, H.J.; *J. Colloid Interface Sci.* **1998**, 198, 241.

10. Weng, L.T.; Bertrand, P.; Stone-Masui, J.H.; Stone, W.E.E.; *Langmuir* **1997**, 13, 2943.

11. Romero-Cano, M.S.; Martin-Rodigez, A.; De-Las-Nieves, F.J.; *J. Colloid Interface Sci.* **2000**, 227, 329.

12. Joanicot, M.; Wong, K.; Richard, J.; Maquet, J.; Cabane, B.; *Macromolecules* **1993**, 26, 3168.

13. Belaroui, F.; Cabane, C.; Dorget, M.; Grohens, Y.; Marie, P.; Holl, Y. *J. Colloid Interface Sci.* **2003**, 262, 336 & 409.

Chapter 5

Predicting Surfactant Distributions in Dried Latex Films

Venkata R. Gundabala[1] and Alexander F. Routh[1,2,*]

[1]Department of Chemical and Process Engineering, University of Sheffield, Mappin Street, Sheffield S1 3JD, United Kingdom
[2]Current address: BP Institute, University of Cambridge, Madingley Road, Cambridge CB3 0EZ, United Kingdom
*Corresponding author: afr10@cam.ac.uk

Abstract

Surfactant transport during the solvent evaporation stage of latex film drying is modeled to predict surfactant distributions across film thickness. Particle and surfactant conservation equations are coupled to take both the surfactant transport due to both particle and molecular diffusion into account. The study shows that if the surfactant adsorption onto particle surfaces is high, the surfactant distribution profiles directly correlate to the particle distribution profiles. The influence of the surfactant adsorption and diffusivity parameters on surfactant distributions obtained at the end of drying is shown. By choosing appropriate adsorption and diffusivity parameters, uniform surfactant distributions can be obtained.

Introduction

Growing regulations on the emission of pollutants from solvent-based coatings has propelled intensive research into their alternatives, the water-based latex coatings. Studies on understanding the mechanism of latex film formation and the factors that control the film forming process, have been the focus of attention in the past several decades. Surfactants are a major ingredient in latex dispersions along with polymer and water. The surfactant distribution, during and at the end of film formation, is found to have a crucial affect on the process of film formation and also on the properties of the dried film. Surfactants, usually added during the latex dispersion synthesis, to provide colloidal stability, affect the particle packing [1], coalescence [2] and polymer chain interdiffusion [3] processes during film formation. Non-uniformities in surfactant distribution in a dried film adversely affect it's adhesive [4] and water-resistant [5] properties. The final properties of latex films are important in a wide variety of industries, and consequently investigation into surfactant distribution has received a great deal of attention in recent years.

Investigations using techniques such as Fourier Transform Infrared (FTIR) Spectroscopy, Atomic Force Microscopy (AFM), Rutherford Backscattering Spectroscopy (RBS), and Raman Spectroscopy have revealed surfactant excesses at both the film-air (F-A) and film-substrate (F-S) interfaces for several surfactant-latex systems under different drying conditions. In the past several years, the aim has been to provide a mechanism for surfactant transport during film drying that will help explain the surfactant non-uniformities that have been observed. The ability of surfactant to reduce interfacial energies has been identified as one of the main driving force for surfactant transport to the interfaces [6]. This observation was corroborated for the case of substrate side enrichment in experiments by Evanson et. al. [7] using FTIR spectroscopy in the attenuated mode. Their studies revealed that low energy substrates always induced greater enrichment towards the substrate side compared to normal substrates. However, the huge excesses sometimes observed at the substrate side cannot be entirely explained by this mechanism alone.

Zhao et. al. [6] suggested surfactant-latex compatibility as a major factor affecting surfactant transport during film drying. This hypothesis was corroborated by other research groups such as Evanson et. al. [8] and Zhao et. al. [9]. While Evanson et. al. used ethyl acrylate/methyl methacrylate latex dispersions and non-ionic surfactants to show that better surfactant-latex compatibility results in lesser exudation towards interfaces, Zhao et. al. showed that lower compatibility between styrene and anionic surfactant was the cause of higher exudations. Kientz et. al. [10] added that the moment of surfactant exudation, a factor affecting the surfactant mobility, is also an important parameter in determining the final distribution of the surfactant. They concluded that exudation occurring late during the film formation process will little affect the final distribution of the surfactant. This was used as a presumption by Gundabala et. al. [11] in an attempt to model surfactant transport during film drying. The model was based on the presumption that surfactant distribution is well established during the solvent evaporation stage and changes little there after. Gundabala et. al. [11] show surfactant adsorption parameters along with a surfactant Peclet number to be the determining parameters. The surfactant conservation equation was solved in the limit of infinite particle Peclet number, Pe_p, the dimensionless number that determines the particle distribution across the film [12]. This limit was invoked by assuming that the speed with which air-water interface moves (evaporation rate) is much higher than the particles can diffuse, leading to dense particle distribution in the upper regions and diffuse distribution in the lower regions. This assumption is acceptable, as the skin formation associated with large Pe_p has been observed by several researchers [13]. The particle Peclet number Pe_p is defined as H_0 / D_p, D_p being the particle diffusivity, and is the only dimensionless group determining how the particles distribute during drying [14]. The particle diffusivity D_p is dependent on the particle size and dispersion viscosity given by the Stokes-Einstein relation. Typical values of Pe_p can vary anywhere from 1 to 100, depending on the drying rate and initial film thickness.

Equally interesting is the generalized case, when the particle Peclet number is not infinite. In this paper, we solve the surfactant conservation equation for the generalized case of any particle Peclet number (relaxing the infinite particle Peclet number limit) in conjunction with the particle conservation equation. This case occurs for slow drying thin films. Routh et. al. [14] solved the particle conservation equation for a drying latex film in the absence of surfactant. We use this equation to obtain a generalized solution to the vertical surfactant distribution problem. In section 2 we derive the surfactant and particle conservation equations and solve them numerically. In section 3 we use the solution to the model to show the influence the particle Peclet number has on the final surfactant distribution and also show the effect of surfactant adsorption parameters on surfactant distribution.

Model

In arriving at the particle conservation equation for a drying latex film, we use a similar approach as in our previous publication [14]. The particle conservation equation [15] is written as

$$\frac{\partial \phi}{\partial \bar{t}} = \frac{1}{Pe_p} \frac{\partial \left(K(\phi) \dfrac{d(\phi Z(\phi))}{d\phi} \dfrac{\partial \phi}{\partial \bar{y}} \right)}{\partial \bar{y}} \qquad (1)$$

Here ϕ is the particle volume fraction and the terms with overbars indicate dimensionless quantities. $\bar{y}$ is the scaled distance from the substrate, distances being scaled with initial film thickness, H_0. $\bar{t}$ is the dimensionless time scaled with characteristic drying time, H_0 / E. E is the fixed velocity of the top air-water interface obtained by assuming a constant evaporation rate from the top surface. $K(\phi)$ is the sedimentation coefficient which has a functional form of $K(\phi) = (1 - \phi)^{6.55}$ [15]. $Z(\phi)$ is the dispersion compressibility taking the form, $Z(\phi) = 1/(\phi_m - \phi)$, where ϕ_m is the closed packed volume fraction. Figure 1 shows a drying film with a receding top surface.

The boundary conditions on equation (1) are no particle flux at the top air-water interface and the bottom substrate side.

$$\bar{y} = 0 \qquad \frac{\partial \phi}{\partial \bar{y}} = 0 \tag{2}$$

$$\bar{y} = 1 - \bar{t} \qquad K(\phi) \frac{d(\phi Z(\phi))}{d\phi} \frac{\partial \phi}{\partial \bar{y}} = Pe_p \phi \tag{3}$$

The surfactant conservation equation is written as

$$\frac{\partial \left(\overline{C_s}(1-\phi) + \overline{\Gamma}\phi \right)}{\partial \bar{t}} = \frac{1}{Pe_s} \frac{\partial \left((1-\phi) \frac{\partial \overline{C_s}}{\partial \bar{y}} \right)}{\partial \bar{y}} + \frac{1}{Pe_p} \frac{\partial \left(K(\phi) \frac{d(\phi Z(\phi))}{d\phi} \overline{\Gamma} \frac{\partial \phi}{\partial \bar{y}} \right)}{\partial \bar{y}} \tag{4}$$

The term on the left hand side is a surfactant accumulation term. The first term on the right hand side is the surfactant diffusion through water and the second term is the surfactant transport through diffusion of particles. The surfactant that is added to a latex dispersion partly remains in solution and partly adsorbs onto the particles based on an equilibrium relation. $\overline{C_s}$ and $\overline{\Gamma}$ are surfactant concentrations in latex serum and on particles, respectively, scaled with the initial surfactant concentration in the latex serum, C_{s0}. Pe_s is the

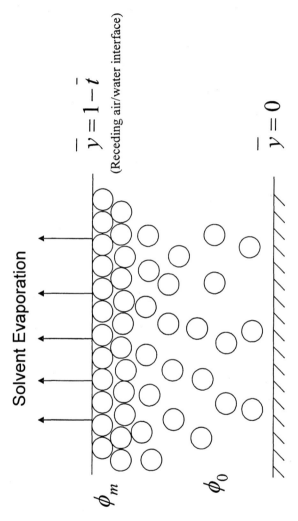

Figure 1. Sketch of a drying latex film with a receding air/water interface at the top

surfactant diffusion Peclet number, defined as $H_0 E / D_S$ giving a measure of the diffusive strength of the surfactant through the latex serum, where D_s is the surfactant diffusion coefficient. Typical Pe_S values are at least an order of magnitude less than Pe_p values and are generally less than 1 under typical drying conditions. The boundary conditions on equation (4) are no surfactant flux at the top and bottom boundaries.

$$\bar{y} = 0 \qquad \frac{\partial \overline{C_S}}{\partial y} = 0 \tag{5}$$

$$\bar{y} = 1 - \bar{t} \quad \frac{1}{Pe_S}(1-\phi)\frac{\partial \overline{C_S}}{\partial y} + \frac{1}{Pe_p}K(\phi)\frac{d(\phi Z(\phi))}{d\phi}\overline{\Gamma}\frac{\partial \phi}{\partial y} = \left((1-\phi)\overline{C_S} + \phi\overline{\Gamma} \right) \tag{6}$$

In order to obtain time independent boundary condition at the top, a new vertical co-ordinate system is defined as $\xi = \dfrac{\bar{y}}{1-\bar{t}}$. This fixes the top surface. The new time variable is $\tau = \bar{t}$

Applying this transformation the new particle conservation equation is

$$\frac{\partial \phi}{\partial \tau} + \frac{\xi}{1-\tau}\frac{\partial \phi}{\partial \xi} = \frac{1}{Pe_p(1-\tau)^2}\frac{\partial\left(K(\phi)\dfrac{d(\phi Z(\phi))}{d\phi}\dfrac{\partial \phi}{\partial \xi} \right)}{\partial \xi} \tag{7}$$

The new conservation for the surfactant is now written as

$$\left(1 - \phi + \phi\frac{d\overline{\Gamma}}{d\overline{C_S}} \right)\left(\frac{\partial \overline{C_S}}{\partial \tau} + \frac{\xi}{1-\tau}\frac{\partial \overline{C_S}}{\partial \xi} \right) + \left(\overline{\Gamma} - \overline{C_S} \right)\left(\frac{\partial \phi}{\partial \tau} + \frac{\xi}{1-\tau}\frac{\partial \phi}{\partial \xi} \right)$$

$$- \frac{1}{Pe_S(1-\tau)^2}\frac{\partial\left((1-\phi)\dfrac{\partial \overline{C_S}}{\partial \xi} \right)}{\partial \xi} - \frac{1}{Pe_p(1-\tau)^2}\frac{\partial\left(K(\phi)\dfrac{d(\phi Z(\phi))}{d\phi}\overline{\Gamma}\dfrac{\partial \phi}{\partial \xi} \right)}{\partial \xi} = 0 \tag{8}$$

The transformed boundary conditions on equation (7) are

$$\xi = 0 \qquad \frac{\partial \phi}{\partial \xi} = 0 \tag{9}$$

$$\xi = 1 \qquad K(\phi) \frac{d(\phi Z(\phi))}{d\phi} \frac{\partial \phi}{\partial \xi} = Pe_p \phi(1 - \tau) \tag{10}$$

The transformed boundary conditions on equation (8) are

$$\xi = 0 \qquad \frac{\partial \overline{C_S}}{\partial \xi} = 0 \tag{11}$$

$$\xi = 1 \quad \frac{1}{Pe_S}(1-\phi)\frac{\partial \overline{C_S}}{\partial \xi} + \frac{1}{Pe_p} K(\phi) \frac{d(\phi Z(\phi))}{d\phi} \overline{\Gamma} \frac{\partial \phi}{\partial \xi} = \left((1-\phi)\overline{C_S} + \phi\overline{\Gamma} \right)(1-\tau) \tag{12}$$

In order to solve the two equations (7) and (8) simultaneously, to obtain the surface distribution across the film thickness, a relation between the surfactant concentrations on the particles, $\overline{\Gamma}$, and in the latex serum, $\overline{C_S}$, is required. By assuming the surfactant adsorption onto latex particles to be of Langmuir type and scaling the particle surface adsorption with initial concentration in the latex serum, the relation below is obtained

$$\overline{\Gamma} = \frac{\dfrac{3\Gamma_\infty}{RC_{S0}}\overline{C_S}}{\dfrac{A}{C_{S0}} + \overline{C_S}} \equiv \frac{k\overline{C_S}}{\bar{A} + \overline{C_S}} \tag{13}$$

Where Γ_∞ is the maximum adsorption onto particles expressed in mol/m^2. A, has units of mol/m^3, giving an estimate of the concentration in the latex serum at which half of maximum particle surface adsorption takes place. These are the Langmuir parameters and k and $\bar{A}$ are their scaled counterparts. It should be noted that in these simulations, the adsorption was taken to be of Langmuir type for simplicity. As shown in a previous work [11], the adsorption may not always be of simple Langmuir type. But for comparison of these model predictions with any experimental results, the relation shown in equation (13) can be easily replaced with the adsorption relation for the particular surfactant-latex system.

Substituting equation (13) into (8), we solve numerically equations (7) and (8) with boundary conditions (9), (10), (11), and (12) using FemLab. FemLab simulations are run in the coefficient mode by defining two variables for $\overline{C_S}$ and ϕ. Discretizing space with 2000 elements and using a time step of 0.001, we obtain the solutions for $\overline{C_S}$ and ϕ as functions of space (distance from substrate) and time. The accuracy of the solution is tested by checking the conservation of surfactant mass.

Results and Discussion

The effect the particle Peclet number, Pe_p, has on the surfactant distribution profile obtained at the end of water evaporation stage is shown in Figure 2. The particle Peclet number Pe_p is varied from 0.5 to 6 and the other system parameters, k, $\bar{A}$, and Pe_S, are kept constant. The % excess/depletion of the surfactant is plotted against the scaled distance from the substrate. The surfactant % excess/depletion at a given distance from the substrate gives a measure of the extent of surfactant enrichment or depletion over the surfactant amount if it were uniformly distributed.

% excess/depletion = (Surfactant amount at that point − Surfactant amount if uniformly distributed)*100

Surfactant amount if uniformly distributed

From Fig. 2, we can see that for all the values of Pe_p, the upper regions of the film are surfactant enriched and the lower regions are surfactant depleted. This is expected and was explained previously [11]. For the operating parameters (k and $\bar{A}$) used here, there is much greater surfactant amount on the particles than in the latex serum, and because the upper regions of the film have higher particle volume fractions compared to the lower regions throughout the drying process, there is greater accumulation in the top region of the film. A direct consequence is the depletion in the lower regions due to surfactant conservation. As Pe_p is decreased from 6 to 0.5, the depletion in the lower regions decreases significantly. As stated earlier, Pe_p gives a measure of the magnitude of particle diffusion. A lower Pe_p indicates greater particle diffusion and hence higher particle volume fractions in the lower regions than for the case of higher Pe_p. So, as Pe_p decreases, the particle volume fraction increases in

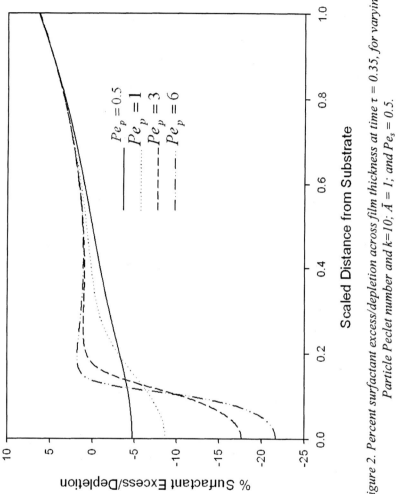

Figure 2. Percent surfactant excess/depletion across film thickness at time τ = 0.35, for varying Particle Peclet number and k=10; Ā = 1; and Pe_s = 0.5.

the lower regions, and because there is considerable surfactant on the particles, the surfactant depletion also decreases.

In an effort to show the effect of Pe_p on the uniformity of the surfactant distribution obtained at the end of drying, we define a parameter Δ, called the degree of non-uniformity, defined as

Δ = Maximum % surfactant excess in the film – Maximum % surfactant depletion in the film.

Fig. 3, shows Δ plotted against Pe_p. As can be seen, Δ increases monotonically with Pe_p, as is expected for surfactant laden particles accumulating at the top surface.

Fig. 4 shows the surfactant distribution profiles obtained by varying k and fixing $\bar{A}$, Pe_s, and Pe_p. A lower k means that there is less surfactant on the particles and a greater amount in the latex serum. The upper regions of the film are enriched (even though not hugely) due to the higher particle volume fractions in the upper regions. The major effect of varying k can be seen in the lower regions of the film. If the k values are sufficiently low, the lower regions are enriched with surfactant and if the k values are sufficiently high, there is depletion in the lower regions of the film. When the k value is low, there is greater surfactant amount in the latex serum and because the water volume fraction is high in the lower regions of the film, surfactant accumulates in this region during the drying process resulting in its enrichment. This observation is similar to the one previously made for infinite Pe_p [11], but here the effect is more pronounced. From Fig. 4, we can also conclude that by choosing appropriate operating parameters (for example, k =3), we can obtain uniform surfactant distributions at the end of water evaporation stage.

Conclusions

This work has modeled the surfactant transport during the solvent evaporation stage of a drying latex film by coupling the particle and surfactant conservation equations. This gives the solution for surfactant distribution for the generalized case of any particle Peclet number. The parametric analysis reveals that apart from the surfactant adsorption parameters and surfactant diffusivity through latex serum, the diffusivity of particles also has an effect on the surfactant distribution profile obtained at the end of water evaporation stage. If the surfactant amount on the particles is considerable, particle distribution through the film thickness greatly affects the surfactant distribution. The extent of the influence of surfactant adsorption parameters on its distribution is also

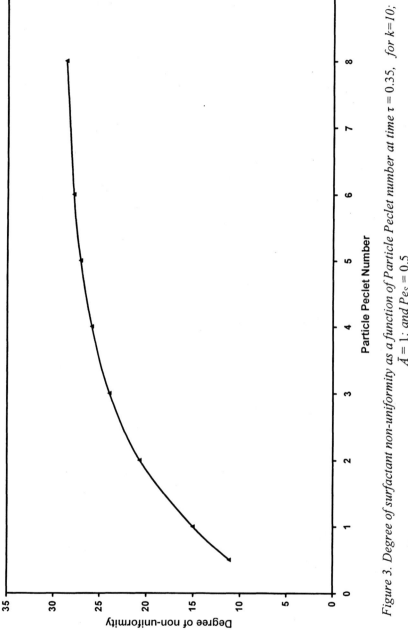

Figure 3. Degree of surfactant non-uniformity as a function of Particle Peclet number at time $\tau = 0.35$, for $k=10$; $\bar{A} = 1$; and $Pe_S = 0.5$

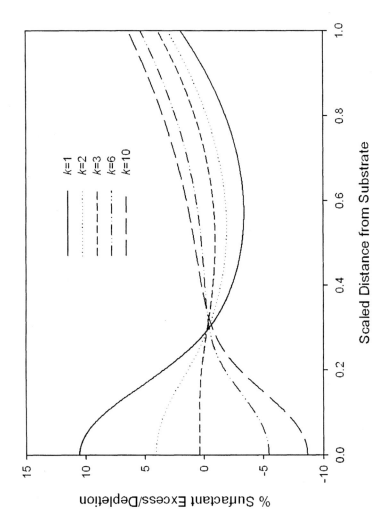

Figure 4. Percent surfactant excess/depletion across film thickness at time $\tau = 0.35$, for varying k values and $Pe_p = 1$; $\bar{A} = 1$; and $Pe_S = 0.5$

largely dependent on the distribution of particles. By adjusting the operating parameters, uniform surfactant distributions can be obtained.

References

1. Okubo, M.; Takeya, T.; Tsutsumi, Y.; Kadooka, T.; Matsumoto, T. *J. Polymer Sci.: Polym. Chem.* **1981**, 19, 1.
2. Eckersley, S.T.; Rudin, A. *J. Appl. Polym. Sci.* **1993**, 48, 1369.
3. Farinha, P.S.; Martinho, J.M.G.; Kawaguchi, S.; Yekta, A.; Winnik, M.A. *J. Phys. Chem.* **1996**, 100, 12552.
4. Zhao, C.L.; Holl, Y.; Pith, T.; Lambla, M. *Br. Polym. J.* **1989**, 21, 155.
5. Mulvihill, J.; Toussaint, A.; De Wilde, M. *Prog. Org. Coat.* **1997**, 30, 127.
6. Zhao, C.L.; Dobler, F.; Pith, T.; Holl, Y.; Lambla, M. *J. Colloid and Interface Sci.* **1989**, 128 (2), 437.
7. Evanson, K.W.; Urban, M.W. *J. Appl. Polym. Sci.* **1991**, 42, 2309.
8. Evanson, K.W.; Thorstenson, T.A.; Urban, M.W. *J. Appl. Polym. Sci.* **1991**, 42, 2297.
9. Zhao, Y.; Urban, M.W. *Macromolecules* **2000**, 33, 2184.
10. Kientz, E.; Holl, Y. Colloids *and Surfaces A: Physiochemical and Engineering Aspects* **1993**, 78, 255.
11. Gundabala, V.R.; Zimmerman, W.B.; Routh, A.F. *Langmuir* **2004**, 20, 8721.
12. Routh, A.F.; Russel, W.B. *A.I.Ch.E Journal* **1998**, 44 (9), 2088.
13. Croll, S.G. *J. Coatings Techn.* **1986**, 58, 41.
14. Routh, A.F.; Zimmerman, W.B. *Chem. Eng. Sci.* **2004**, 59, 2961.
15. Russel, W.B.; Saville, D.A.; Schowalter, W.R. *Colloidal Dispersions*; Cambridge University Press: Cambridge, 1995.

Morphology Characterization, Novel Morphologies, and Film Structures

Chapter 6

Cryo-Scanning Electron Microscopy of Film Formation in Regular and Low-Volatile Organic Contents Waterborne Latex Coatings

Haiyan Ge[1], Cheng-Le Zhao[2], Shane Porzio[2], Li Zhuo[2], H. T. Davis[1], and L. E. Scriven[1]

[1]Department of Chemical Engineering and Materials Science, University of Minnesota, Minneapolis, MN 55455
[2]BASF Charlotte Technical Center, 11501 Steele Creek Road, Charlotte, NC 28273

The film formation processes in regular and low-VOC waterborne latex coatings were compared by cryogenic scanning electron microscopy (Cryo-SEM). Five model MMA/BA/AA/AM copolymer dispersions with particle diameters around 130 nm were used. They were coated on 5mm by 7mm silicon wafers and after 1, 10, and 60 minutes of drying were rapidly frozen in liquid ethane, fractured, sublimated to develop contrast, coated to avoid charging, and examined on the cold stage. Images of the fracture surfaces show top-down and edge-in gradients of consolidation, compaction and coalescence. Some of the low-VOC coatings form a film as fast as conventional ones—or even faster. The findings are corroborated by results of SEM and scanning probe microscopy (SPM) of fracture surfaces and top surfaces of coatings dried for 10 days. Cryo-SEM was also used to compare the degree of swelling and particle shape recovery when the dried coatings were re-hydrated.

69

Cryogenic scanning electron microscopy (Cryo-SEM) of latex coatings frozen at successive stages of microstructure evolution, fractured to expose cross-sections, and imaged at high resolution has revealed unprecedented detail of the film-formation process[1-4]. The three stages of latex film formation are:1) consolidation – solvent evaporation and particle packing; 2) compaction – particle deformation and pore-space shrinkage; and 3) coalescence – particle adhesion and interparticle diffusion and entanglement of polymer. These can be consecutive or overlapping. Different stages may coexist in different parts of drying coatings. Away from the edges, they proceed from the exposed top surface of the coating to its bottom; whereas near the edge, where there is more air accessible per unit area of surface and drying is generally faster, they proceed from the edge inward, i.e. toward the central area of the coating. Thus the process inherently involves a gradient perpendicular to the substrate, i.e. top-down gradient, and elsewhere a gradient nearly parallel to the substrate, i.e. edge-in gradient. These gradients may give rise to defects in the dried coating, depending on the physical properties of the latex and the drying rate. For example, a skin may become evident during the drying of a latex that is soft enough. The greater shrinkage and higher modulus of a skin can produce in-plane residual tensile stress great enough to cause unwanted curling or even cracking of industrial coatings.

Plasticizing organic solvents like TPM (2,2,4-trimethyl-1,3-pentanediol monoisobutyrate with the trade name Texanol[TM] from Eastman Chemical Company) have been commonly used to lower polymer glass transition temperature (T_g) and thereby to speed and enhance compaction and coalescence. Lowering the amount of such volatile organic compounds (VOC) in water-based latex coatings without loss of mechanical strength and other properties is a challenge to both technological development and scientific understanding[5-12]. Three leading strategies of lowering VOC content are 1) blending particles of a "film former", i.e. low T_g polymer with particles of a "non-film former", i.e. high T_g polymer[4,6-7]; 2) developing composite latex particles that contain two or more polymers[9]; 3) low-T_g latex particles ($T_g < 25°C$) containing a room temperature crosslinking system[10-12]. Perhaps because of the lack of adequate experimental and analytical techniques, most of the studies of low-VOC latex coatings published over the past decade have been carried out on dried coatings. It therefore has not been possible to elucidate the drying mechanism, let alone understand how the drying process itself affects the properties of dried coatings.

In this paper, we reported a study of the film formation process encompassing all three stages with an aim to better understand the basic principles underlining the three low-VOC strategies. How film formation proceeds in time was probed by Cryo-SEM and supported by scanning probe microscopy (SPM). We examined the influence of a coating's nominal Minimum Film Formation Temperature (MFFT); the presence of the common coalescing aid TPM; the particles' initial morphology; and the use of a room-temperature crosslinking chemistry.

Materials and Methods

Latex

Five acrylic copolymer model latexes were synthesized by a seeded semi-continuous emulsion polymerization process. A fixed amount of fine seed particles was used for all polymerizations in order to control particle size. Monomers (see Table 1) were prep-emulsified using SDS (Sodium Dodecyl Sulfate, 98% from Sigma-Aldrich) and water. The monomer emulsion and initiator solution (ammonium persulfate in water) were fed separately and continuously into the reactor under "starving conditions". Except for the surfactant SDS, all other chemicals used in this work were of industrial grade. The compositions of the latexes are listed in Table 1. After emulsion polymerization, the latexes were neutralized with concentrated ammonia to pH values around 8.5. The five samples were designed to compare effects of high and low-MFFT's, a room temperature crosslinking system, and a hard-soft particle morphology. Their solid content, MFFT and particle diameter are shown in Table 2.

Cryo-SEM of Drying Latex Coatings

The polymer latexes were coated on 5mm by 7mm silicon wafers with a wire-wound rod, air-dried for various periods at 21°C to 23 °C and 50% relative humidity, and fast-frozen by hand-plunging in liquid ethane. The frozen specimens were loaded onto a sample holder (developed by Sutanto et al.[2]) in a liquid nitrogen bath and then transferred to an Emitech K-1250 Cryo-system (Empdirect, Houston, TX) pre-cooled to liquid nitrogen temperature. The specimens were fractured and the cross-sections were sublimated for 9 minutes at - 96°C to develop topographic contrast. They were then sputtered at -120°C

Table 1. Composition of The Five Copolymer Latexes

Proportions by weight	MMA	BA	AA	AM	DAAM
SF01	51	46.5	1.5	1	0
SF01B	51	46.5	1.5	1	0
SF02	37.5	60	1.5	1	0
SF03	35.5	60	1.5	1	2
SF04	Proprietary				

*MMA–methyl methacrylate; BA–n-butyl acrylate; AA–acrylic acid; AM–acrylamide; DAAM–diacetone acrylamide; all latexes contain 1 wt % of sodium dodecyl surfate and 0.2 wt % of ammonium persulfate (based on dry polymer weight).

Table 2. Key Features of the Five Copolymer Latexes

	Polymer Characteristics	Solids Content wt.%	MFFT °C	Diameter nm
SF01	Conventional high MFFT	50.8	17	129
SF01B	SF01 with 10 wt% coalescing aid Texanol^{TM}	50.8	0	129
SF02	Low MFFT	50.6	0	129
SF03	Low MFFT and crosslinkable polymer	47.8	0	129
SF04	Low MFFT and hard-soft two-phase particles	50.0	0	120

* 10 wt % of TPM (based on dry polymer weight) was added to SF01; 1 wt % adipic dihydrazide ADDH (based on dry polymer weight) was added to SF03.

with a layer of platinum around 2-3 nm thick and transferred into a Hitachi S4700 below-the-lens Field Emission SEM for examination at a low acceleration voltage of 2 KeV and a low probe current of about 10^{-11} A. During the imaging, the sample temperature was kept around -160 °C.

Room temperature SEM images of cross-sections of latex coatings, which were dried for 10 days and then fractured in liquid nitrogen, were made with the Hitachi S4700. Some of the dried coatings were soaked in deionized water for 1 day as a means of evaluating water sensitivity. Particle swelling and shape recovery were examined by observing both their visual appearance and by Cryo-SEM, their cross-sections.

Tapping Mode SPM of Dried Latex Coating Surfaces

Top surfaces of samples prepared in the same way and dried for 10 days were imaged with a Nanoscope III Multimode scanning probe microscope (Digital Instrument Co., Santa Barbara, CA) operated in the Tapping Mode at ambient conditions. To preclude any effect of surfactant and other water-soluble substances that might be squeezed out during film formation to the top surface, the same dried samples were placed in deionized water for 5 minutes, washed by hand shaking for 2 minutes, allowed to dry for 24 hours, and reexamined by SPM. The images made before washing did not differ perceptibly from those after washing and so are not shown[13].

Results and Discussion

Top-Down and Edge-In Gradients in Drying Latex Coatings

The top-down and edge-in gradients are evident in micrographs of each of the model latex systems. A good example of the development of these gradients during drying is shown in the two time sections of SF01 latex coating. Figure 1 shows freeze fracture Cryo-SEM images of SF01 latex coatings dried for 1 minute and 10 minutes. The images were captured in the central area of each coating. The images on the left of Figure 1 illustrate the degree of consolidation after 1 minute of drying. The particles have begun contacting with each other, but the dimples in the ice induced by plucking out of latex particles during freeze fracture show that there is still enough water to separate particles in some places. There appears to be progressively more water remaining between particles from top down. The images on the right of Figure 1 show that after 10 minutes of drying, the particles are consolidated throughout the coating depth and have begun compacting. Those in and near the top surface appear more flattened against each other than those deep down. Nevertheless, the top-down gradient is less steep than at 1 minute of dying time and the coating is becoming more uniform.

Figure 2 shows a sequence of time-sections by Cryo-SEM of SF01 latex coatings dried for 1, 10 and 60 minutes. These images were captured at mid-depth near the edge and in the central area of each coating; images of other depths are not shown here[13]. Figures 2 a), c) and e) are from near the edge and Figures 2 b), d) and f) are from the central area. At 1 minute of drying, particles near the edge are consolidated, whereas those in the middle are still surrounded by frozen water. At 10 minutes of drying, particles near the edge are compacted, i.e. they are flattened against each other into polyhedral shape, whereas those in the middle of the coating are merely consolidated or have only begun to compact. At 1 hour of drying, the difference between edge and center has become small: the coating is uniformly compacted. There are features in both Figures 1 and 2 that appear brighter than their surroundings on the facture surfaces. These are called "pullouts" because they are partially elongated particles that remain in the fracture surface, as many other images of inclined fractures and tilted specimens show. Here, their three-dimensional aspect is most evident in Figures 1 c) and 2 e). Pullouts are generated by plastic deformation of particles during the freeze-fracture step. Their presence and appearance have proven useful indicators of the local stage of film formation in latex coatings[14].

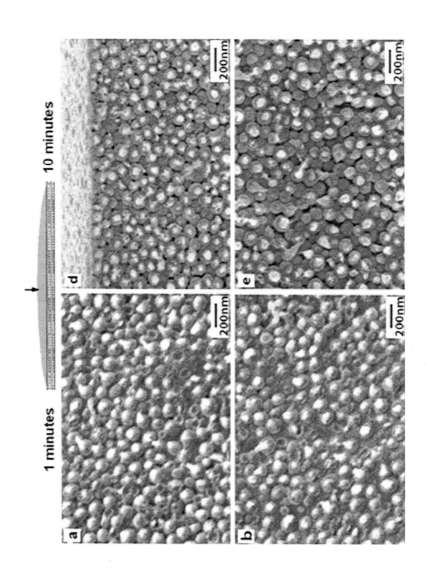

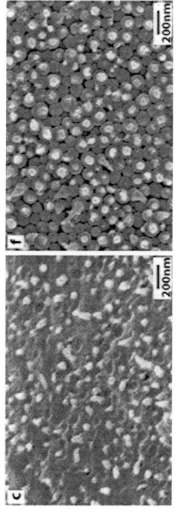

Figure 1. Top-down gradients in SF01 coatings dried for 1 minute a) near top surface, b) in the middle of coating, c) near bottom, and for 10 minutes d) near top surface, e) in the middle of coating, f) near bottom.

76

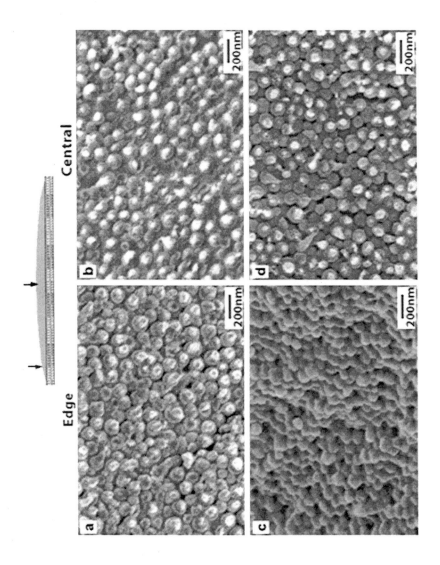

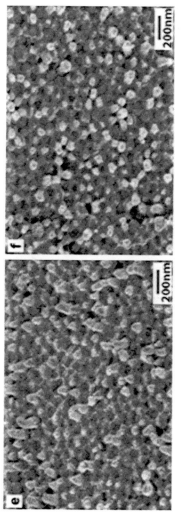

Figure 2. Edge-In gradients in SF01 coatings dried for 1 minute a) near edge, b) in the center, for 10 minutes c) near edge, d) in the center, for 60 minutes e) near edge, f) in the center.

Because drying of the coating is generally faster near the edge, than far away from it, the gradient from the edge toward the central area is more evident than that from the top surface toward the substrate; with time the edge-in gradient diminishes more slowly that the top-down one. No abrupt consolidation or compaction front was observed because the model latex systems were soft enough that the drying stages overlapped (in contrast to cases of high Tg latexes examined by Sutanto et al.[2]). For the purpose of comparison, the freeze-fracture Cryo-SEM images in the following figures are from the mid-depth in the central area of each coating.

Effects of Coalescing Aid

Figure 3 shows two time sequences, one of coatings with coalescing aid (SF01B) and the other, without (SF01). The Cryo-SEM images are of fracture surfaces of coatings dried for 1, 10, and 60 minutes; the SEM images are of coatings dried for 10 days. At 1 minute of drying, both coatings are still consolidating. At 10 minutes of drying, particles in the coating with coalescing aid appear to have fused, because their boundaries cannot be discerned and the fracture surface has diagnostic pullout features longer than the original particles' diameter. Particles in the coating without coalescing aid are still distinct and incompletely consolidated, although some have begun compacting. At 60 minutes of drying, the coating with coalescing aid has long since coalesced. In comparison, particles in the coating without coalescing aid are compacted, but still discernible because most of the pullouts are from individual particles. At 10 days of drying, the fracture surface of the coating with coalescing aid is smooth, whereas that of the coating without coalescing aid appears rough and some particles still can be discerned. This evidence supports the hypothesis that coalescing aids promote film formation by lowering the elastic modulus and yield strength of the particles so that they flatten against each other faster; the evidence also buttresses findings that film formation is promoted by raising the mobility of the polymer chains and their rate of diffusion across the enlarging contact surfaces[15]. However, coalescing aids raise the level of VOC in latex paints

Effects of Low MFFT

Figure 4 and Figure 5 show time sequences of Cryo-SEM images of the three low MFFT coatings, SF02, SF03 and SF04, dried for 1, 10 and 60 minutes, plus SEM images of the three coatings dried for 10 days. Comparing the low MFFT coatings SF02, SF03 and SF04 of Figures 4 and 5 with the high. MFFT coating in Figure 3, all of which were dried under the same conditions, brings out that the particles in the low MFFT coatings compacted and coalesced more rapidly. What is remarkable is that the low MFFT coating SF03 which contains a small amount of water-soluble crosslinker, appears to coalesce faster than those without; after 10 minutes of drying, the SF03 certainly looks like it is ahead; in comparison, both SF02 and SF04 coatings are still consolidated. After 10 days of drying, all three of the low MFFT coatings are coalesced, as judged by their smooth fracture surfaces on which there is almost no contrast: observe Figure 4 d) and h) and Figure 5 d).

Figure 6 shows SPM images, or "3D surface maps," of the top surfaces of the full suite of five coatings after they dried for 10 days and then were washed and redried for 1 day. The surface of the high MFFT SF01 latex coating has higher root-mean-square roughness (RMS) than the others. SPM has better resolving power than SEM in the "z-direction," down to a few angstrom in some cases. The SPM images of top surfaces show that leveling is still incomplete at 10 days; whereas the SEM images of fracture surfaces show no discernible difference, with the exception of the SF01 latex coating. Thus a coating may be coalesced even though the top surface is not fully leveled according to SPM. In other words, SPM of the degree of leveling of the top surface may be an overconservative indicator of coalescence.

With the exception of the dried SF04 coating, the SPM images reveal clearly the outlines of the latex particles at top surfaces (Figure 6). In the case of the hard-soft composite SF04, the characteristic length scale is longer and variation at the particle scale is muted, suggesting that the softer material has coalesced, leaving a "plum pudding" of hard domains. The phase image (in contrast to height image) of SF04 coating confirms this inference, which shows two materials with different stiffness dispersing on the top surface[13].

Dried Coatings After Rewetting

After rewetting for a day, the high MFFT SF01 coating appears visually opaque; the SF01B and SF02 coatings appear transparent; SF03 and SF04 coatings, opaque. These appearances accord with the freeze-fracture Cryo-SEM images of the soaked coatings in Figure 7, as follows. The coalescence of SF01 coating evidently was not enough to prevent water from diffusing in and releasing at least some of the particles to recover to their original spherical

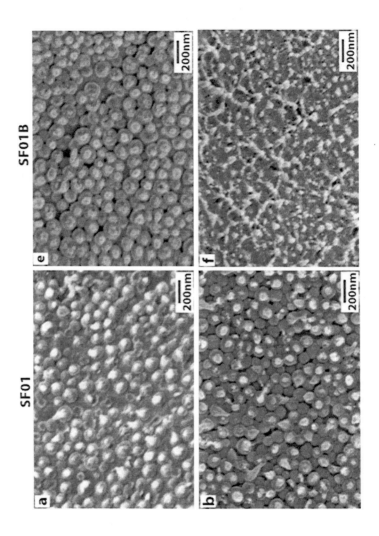

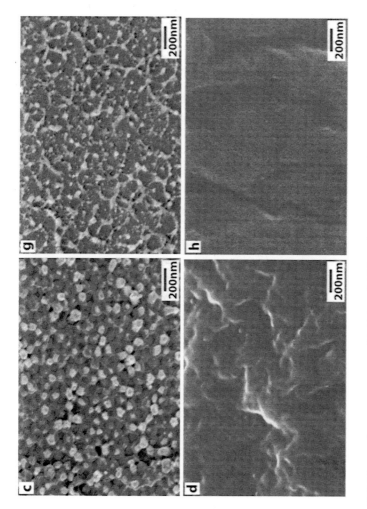

Figure 3. Cryo-SEM freeze fracture images of SF01 and SF01B latex coatings dried a) and e) 1 minute; b) and f) 10 minutes; c) and g) 60 minutes; and d) and h) SEM freeze fracture images of SF01 and SF01B latex coatings dried for 10 days.

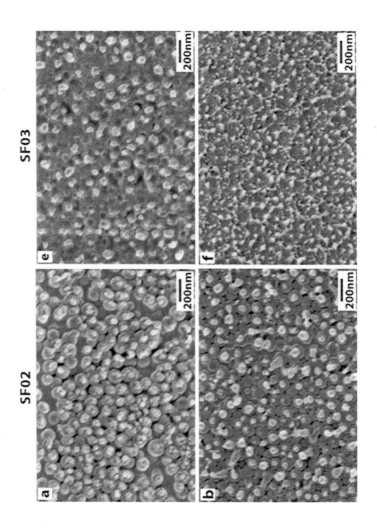

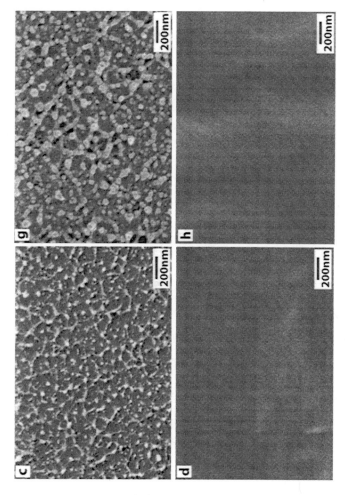

Figure 4. Cryo-SEM freeze fracture images of SF02 and SF03 latex coatings dried a) and e) 1 minute; b) and f) 10 minutes; c) and g) 60 minutes; and d) and h) SEM freeze fracture images of SF02 and SF03 latex coatings dried for 10 days.

84

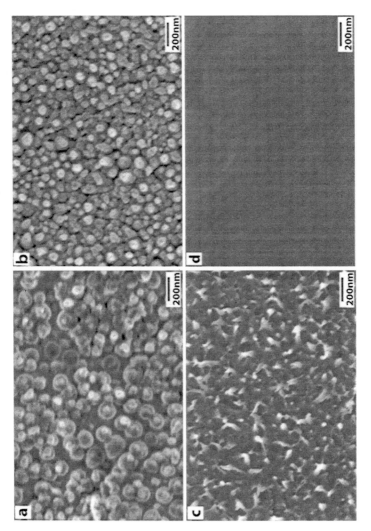

Figure 5. Cryo-SEM freeze fracture of SF04 coating dried for a) 1minute; b) 10 minutes; c) 1hour; and d) SEM freeze fracture of SF04 coating dried for 10 days.

shape, indicating that they had been elastically flattened. The coalescence aid-laden coatings SF01B and the low MFFT SF02 coating evidently developed enough interparticle adhesion, or coalescence, in their compacted states that water could neither release particles from one another nor open enough porespace to scatter much light; hence they remained transparent. The water-soluble crosslinkable polymer in the SF03 coating presumably reacted with functional groups on the particles' surfaces once they were consolidated. Were this crosslinking more rapid than the diffusion of polymer chains across contact surfaces, the diffusion would be arrested; that would leave polar constituents of the system more concentrated in the interparticle boundaries[10-12], where they can facilitate water uptake. Rapid enough crosslinking would account for the SF03's ability to take water up and turn opaque. In the hard-soft composite, SF04, the fracture surface appears porous and rough; microcracks appear to have swelled with water during the soaking, and then the water turned to ice during the plunge freezing; thereafter, some of the ice was sublimated away. It may be that during drying, net in-plane tensile forces between particles developed faster than adhesion and tensile strength grew between many pairs.

Conclusion

Cryo-SEM images document compaction and aspects of adhesion and coalescence as latex film formation proceeds during drying. Rewetting tests can be useful to elucidate incomplete coalescence. Room temperature SEM of fracture surfaces and SPM of top surfaces of dry or nearly dry coatings are useful adjuncts. The top surface of a coalesced coating may still be incompletely leveled. Lowering MFFT of course speeds coalescence ceteris paribus. Lowered VOC coatings with hard-soft initial particle morphology or interparticle crosslinking during compaction can, in some cases, form film as fast as, or faster than comparable conventional compostions. However, these two lowered VOC coatings have poor resistance to water uptake as a result of identifiable features in their film formation processes.

Acknowledgements

This paper reports joint research of the Coating Process Fundamental Program at the University of Minnesota and the Charlotte Technical Center of BASF Corporation The authors are grateful to Scott B. Robinson and Anthony J. Robinson of BASF for their synthesizing and characterizing model latexes.

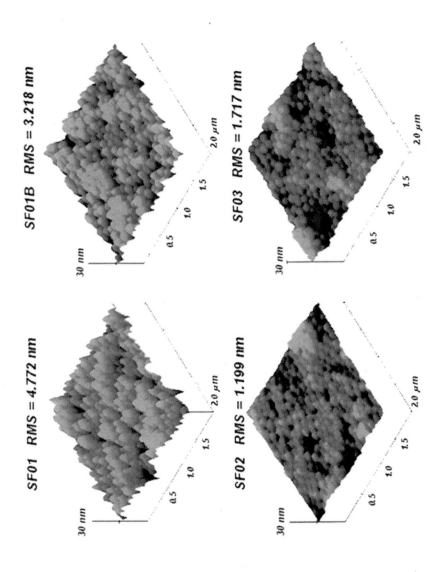

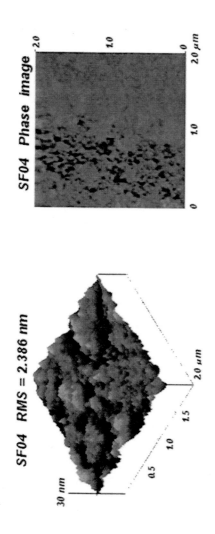

Figure 6. Tapping Mode SPM images of the top surface of the coatings from 5 samples dried for 10 days.

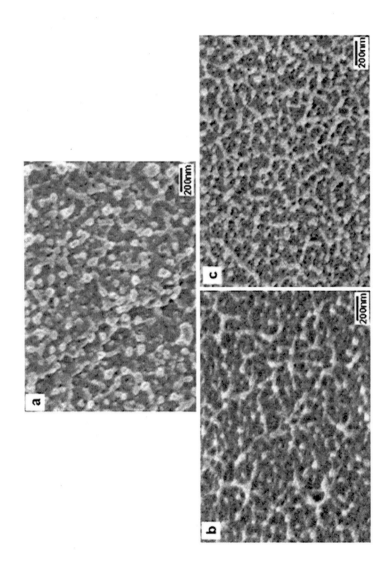

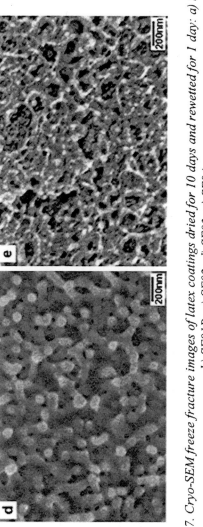

Figure 7. Cryo-SEM freeze fracture images of latex coatings dried for 10 days and rewetted for 1 day: a) SF01; b) SF01B; c) SF02; d) SF03; e) SF04.

References

1. Huang, Z.; Thiagarajan, V. S.; Lyngberg, O. K.; Scriven, L. E.; Flickinger, M. C. *J. Coll. and Interface Sc.* **1999**, *215*, 226-243.

2. Sutanto, E.; Ma, Y.; Dittrich, U.; Zhao, C. L.; Zhuo, L.; Davis, H. T.; Scriven, L. E. *Proceedings of the 10th International Coating Science & Technology Symposium* 2000, pp 63-66.

3. Sutanto, E.; Ma, Y.; Davis, H. T.; Scriven, L. E. *Film Formation in Coatings*; ACS Symposium Series, 790; 2001; pp *174-192*.

4. Ma, Y.; Wiley, B. J.; Davis, H. T.; Scriven, L. E. *Proceedings of the Ann. Meet. Tech. Prog. of the FSCT (2002) 80th*; pp 212-256.

5. Friel, J. M. Eur. Patent 466, 409, 1991.

6. Winnik, M. A.; Feng, J. R. *J. Coatings Tech.* **1996**, *68(852)*, 39-50.

7. Eckerskey, S. T.; Helmer, B. J. *J. Coatings Tech.* **1997**, *69(864)*, 97-107.

8. Wicks, Z. W.; Jones, F. N.; Pappas, S. P. *Organic Coatings-Science and Technology;* Wiley-Interscience, 1999; pp 328-334.

9. Schellenberg, C.; Tauer, K.; Antonietti, M. *Macromolecule Symp.* **2000**, *151*, 465-471.

10. Tronc, F.; Liu, R.; Winnik, M. A.; Eckerskey, S. T.; Rose, G. D.; et. al. *J. Polym. Sci. Part A: Polym. Chem.* **2002**, *40*, 2609-2625.

11. Toussaint, A.; De Wilde, M.; Molenaar, F.; Mulvihill, *Prog. in Organic Coatings.* **1997**, *30*, 179-184.

12. Aradiam, A.; Raphael, E.; de Gennes, P. -G. *Macromolecules* **2002**, *35*, 4036-4043.

13. Ge, H. Ph.D. thesis, University of Minnesota, Minneapolis, MN, 2005.

14. Ge, H.; Davis, H. T.; Scriven, L. E.; Zhao, C.; Prozio, S.; Zhuo, L. *Manuscript in preparation.*

15. Wang, Y. C.; Winnik, M. A. *Macromolecules* **1990**, *23*, 4731-4732.

Chapter 7

Characterization of Free-Volume Properties of Surface and Interfaces in Thin Film Using Positron Annihilation Spectroscopy

Junjie Zhang, Hongmin Chen, Quang Liu, L. Chakka, and Y. C. Jean

Department of Chemistry, University of Missouri at Kansas City, Kansas City, MO 64110

Positron annihilation spectroscopy (PAS) is a special nondestructive evaluation (NDE) technique for materials characterization, which uses the positron (anti-electron) as a radio-analytical probe. This paper presents some of our recent results on the applications of positron annihilation lifetime spectroscopy and Doppler broadening energy spectroscopy coupled with a variable energy positron beam to characterize free-volume properties in nano-scale polymeric films. Depth profiles of glass transition temperature and layered structures in polystyrene thin films on the gold substrate are determined.

Introduction

Positron annihilation spectroscopy (PAS) is a branch of γ-ray spectroscopy which monitors the lifetime, energy spectrum, and angular correlation of annihilation photons [1]. PAS has been used to study defects in solids for many decades. Recently, it has been successfully applied to measure the free-volume properties in polymers [2,3] and in thin films [4-18].

With its unique sensitivity to the atomic-level free volume in polymers, PAS is emerging as a promising tool to measure T_g as a function of depth [7,9,13,14,17,18]. Positron annihilation lifetime (PAL) spectroscopy is capable of determining size, quantity, distribution, and relative fraction of free volume in polymers due to the fact that the ortho-Positronium (o-Ps, the triplet Positronium) is preferentially trapped in the subnanoscale free volume. In this paper, we report the nanoscale structures from the measured free-volume data in thin polystyrene films supported on the Au substrate.

Glass transition temperature T_g in thin films is one of the most fundamental physical properties for their applications to chemical and electronics industries [19]. A variety of techniques, such as Brillouin light scattering [20,21], ellipsometry [22], neutron scattering [23,24], Differential Scanning Calorimetry (DSC) [25,26], Scanning Force Microscopy (SFM) [27], Atomic Force Microscopy (AFM) [28], sum-frequency vibrational spectroscopy [29], and fluorescence spectroscopy [30,31] have been used to measure the T_g of thin polymeric films. However, despite much effort put into this area of research in recent years, existing reported results are still inconsistent and interpretations are not settled. In a recent study on an 80-nm PS film on Si substrate [17], we found that T_g is 17 K lower near the surface and 11 K lower at interfaces than in the bulk. This is different from the earlier study by another group who could not detect a T_g-depth dependence [7]. In this paper we report the depth variation of T_g at the surface and interfaces in a thin polystyrene film on the Au substrate [18].

Experiments

The polystyrene (PS) used in this study was purchased from Aldrich Chemicals (M_w = 212,400, M_w/M_n = 1.06). Gold- (50 ± 3 nm) coated glass slides were from Platypus Technology LLT (Madison, WI) and were cleaned by acetone, rinsed with deionized water, and then dried in vacuum. The polystyrene films were prepared by dissolving 5 wt % polystyrene into toluene, then spin-coated onto Si wafers at a spin rate of 2000 RPM for 5 min. The films were annealed in vacuum at 150 °C for 12 h before mounting on a Kapton film heater in a vacuum system for PAS measurement. The film thickness was found to be

150 ± 5 nm on Au using profilometry (Tencor alfa-step 200, Adv. Surf. Tech., Cleveland, OH) as described elsewhere [18]. The surface roughness of Au was measured to be 3 ± 1 nm.

Doppler broadening energy spectroscopy (DBES) and PAL were measured at the University of Missouri–Kansas City (UMKC) using a slow positron beam with a variable energy from 0 to 30 keV [15,16]. DBES was measured with positron energy from 0.1 keV to 30 keV at a counting rate of 2000 cps [15]. Detailed descriptions of the UMKC positron beam can be found elsewhere [15-18].

The obtained DBES spectrum is expressed as the S parameter, which is defined as a ratio of integrated counts near 511 keV at the central part to the total counts with a window of ±0.53 keV. The S parameter represents a measure of the free-volume quantity in polymeric materials from two main contributions: the para-Positronium (p-Ps, singlet Ps) annihilation and the energy broadening due to the uncertainty principle for Ps and the positron localized in a free-volume hole. The energy resolution of the solid detector was 1.5 keV at 511 keV. One advantage of measuring the S parameter over the positron lifetime is the short time of data acquisition (on the order of minutes) as opposed to hours for a PAL spectrum total of 1 million counts. A detailed description of DBES and the S parameter can be found elsewhere [15-18].

PAL experiments were performed using a 0-30 keV positron beam [15,16]. The PAL data thus contain free-volume properties for polymers from the surface, any interfaces, and to the bulk. The lifetime resolution was 350-500 ps from the bulk to the surface at a counting rate of 100-300 cps. Each PAL spectrum contains 2 million counts. The obtained PAL data were fit into three lifetime components using the PATFIT program [32] after the longest lifetime component (>10 ns) was subtracted from the raw spectra, and the subtracted spectra was then fit in the range to 25 ns as we have reported in the past [33,34]. Each PAL spectrum contains p-Ps, the positron, and o-Ps annihilation radiations in polymers. Fortunately, in complicated polymeric systems, it has been shown that Ps preferentially localizes in defect sites [35], particularly in the free volume before annihilation takes place. Therefore all Ps signals contain the electron properties of free volume. Three resolved lifetimes in polymers are $\tau_1 \sim 0.125$ ns which corresponds to p-Ps annihilation, $\tau_2 \sim 0.45$ ns from the positron annihilation, and τ_3 due to o-Ps annihilation. The o-Ps lifetime τ_3 is relatively longer than the others, on the order of 1-5 ns in polymeric materials, the so-called pickoff annihilation with electrons in molecules [2,3]. A correlation between the measured o-Ps lifetime τ_3 and the free-volume radius R based on a spherical-cavity model [36] has been established as:

$$\tau_3^{-1} = 2\left[1 - \frac{R}{R_0} + \frac{1}{2\pi}\sin\left(\frac{2\pi R}{R_0}\right)\right] \quad (ns^{-1}) \qquad (1)$$

where $R_0 = R + \Delta R$, and ΔR is an empirical parameter determined to be 1.656 Å [37] by fitting the observed lifetimes with the known hole and cavity sizes in molecular substrates. Furthermore, the intensity or the probability of o-Ps lifetime I_3 may be used as information about the relative numbers of free volume [2,37]; then free volume V_f based on R from Eq (1) and I_3 are the foundation for the determination of relative fractional free volume (ffv) [2]. A good correlation has been calibrated for hole sizes up to about a radius of 1 nm. Recently this equation has been extended for hole sizes larger than 1 nm or o-Ps lifetimes longer than 10 ns [38,39]. It is assumed that in the larger pores the o-Ps behaves more like a quantum particle, bouncing back and forth between the energy barriers as the potential well becomes large.

A PAL spectrum in a polymeric material could be resolved into a continuous lifetime distribution since the free volume has a distribution. All PAL spectra were further analyzed into continuous lifetime distributions using three existing programs: CONTIN [40], LT [41], and MELT [42]. While they provide similar results, we only present the smoothed lifetime distributions from MELT analysis here.

When a positron enters the polymeric surface, it loses its energy via inelastic collision processes, which could be expressed by a Makhovian implantation profile. The mean depth Z of the polymer where the positron annihilation occurs is calculated from E_+ using Eq (2), below [43]:

$$Z(E_+) = (40 \times 10^3 / \rho)E_+^{1.6} \qquad (2)$$

where Z is expressed in nm, ρ is the density in kg/m^3 and E_+ is the positron incident energy in keV. The depth resolution is better near the surface than in the bulk. A typical depth resolution is estimated to be about 10% of the depth of interest. All positron beam experiments were performed under high vacuum of 10^{-8} torr.

The polymeric samples were heated using a resistance Kapton device which was controlled by an Omega temperature controller with accuracy of $\pm$ 1°C. The heating process started from r.t. with 5°C intervals to 150°C, then cooled to r.t.

Results and Discussion

Figure 1 shows the variation of the S parameter measured by DBES as a function of positron energy or mean depth (top x-axis) as calculated from the positron energy according to the established Eq (2) for a thick PS film, Au coated on a glass plate, and a thin PS film coated on a Au-glass substrate. A low value of S parameter at low positron energy for the PS films is due to back diffusion of the implanted positron and Ps from the polymer. In a polymer the diffusion length of Ps and positron is small, on the order of 1-10 nm and 100 nm for Ps and the positron, respectively [35], due to trapping in the free volume. On the other hand, for Au-coated glass substrate a decrease of S from the surface to the bulk is due to a small value of S for metallic gold as the positron penetrates to Au and then the glass plate.

The layered structure of 150-nm Ps on Au/glass substrate can be fitted by taking the positron energy distribution as function of depth according to the Makhovian implantation profile using the computer program VEPFIT [44]. We have tried multi-layer structure fitting from three to five layers and found the four-layer fit gives a good fitting result [18]. While VEPFIT program is model dependent, the currently fitted four-layer model is consistent with that previously reported result for an 80 nm PS on the Si substrate [17]. However, the fitted result for PS on Si substrate [17] gave a significantly larger (21 nm) interface than the current PS on Au substrate, i.e. 2 nm interface. Four regions for a polystyrene film are identified: the near surface and the polymeric film (region I), the interfacial layer between the polymer and Au substrate (region II), the Au layer (region III), and the glass substrate (region IV), schematically shown in Figure 2. The resolved film thickness (155 ± 10 nm) from VEPFIT analysis agrees well with the result obtained from profilometry (150 ± 5 nm) [18].

The main VEPFIT results of the S parameter vs depth for a 150-nm PS film are: the density of the interface layer (0.5 ± 0.3 g/cm^3) is significantly lower than that of the film layer (1.1 g/cm^3); the thickness of the film layer and the interface layer is 155 ± 10 nm and 2 ± 1 nm, respectively; the Ps diffusion length is very short, 5 ± 2 nm and 2 ± 1 nm in the film and in the interface, respectively. The S parameter data is thus useful for the determination of layer structures for polymeric thin films. The longer diffusion length in glass and Au is due to little or no Ps formation in glass and Au, respectively. Table 1 summarizes these results.

PAL provides quantitative information about free-volume size and distribution. Figure 3 shows the results of o-Ps lifetime τ_3 and intensity I_3 from PAL spectra vs energy or mean depth in a 150-nm Ps film on Au/glass substrate. The o-Ps lifetime is large at the surface and decreases to 2.05 ns at the film layer. It is interesting to observe that the variation of o-Ps intensity (I_3 of Figure 3) is similar to that of the S parameter in Figure 1. This is expected because S mainly determines p-Ps, while I_3 is the intensity of o-Ps. At the surface, the low intensity is due to backscattering of positron and Positronium. Similar to S data, it increases to its peak value at the film layer, then decreases at the interface layer and substrates

Glass transition temperature T_g can be defined accurately by measuring the molecular level of free-volume variation with respect to temperature. The DSC measures the variation of heat capacity rate with temperature, and the intercept

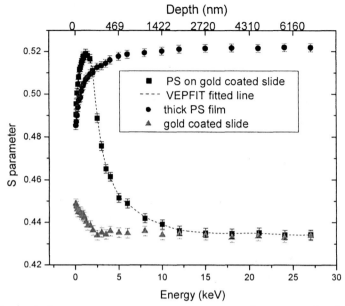

Figure 1. S parameter variation vs. positron incident energy for thick polystyrene film, a gold coated glass slide and thin 150 nm polystyrene film on gold coated glass slide. The line through thin film is from a four-layer model fit using VEPFIT program [44].

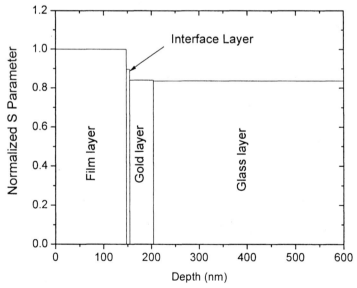

Figure 2. A schematic diagram of four layer structure model from fitted S parameter data for a 150 nm PS film on gold coated glass substrate.

TABLE 1. Fitted result for polystyrene on gold-coated glass.

Layer	S parameter	Density (g/cm^3)	Diffusion length (nm)	Thickness (nm)
Polystyrene	0.505±0.003	1.10	5 ± 2	155±10
Interface	0.461±0.021	0.5 ± 0.3	2±1	2 ± 1
Gold	0.435±0.004	19.3	18±3	50
Glass slide	0.433±0.012	2.6	19±3	substrate

point was found to be $T_g = 100 \pm 1$ °C for a thick PS film [8]. We measured the free-volume size (from o-Ps lifetime or τ_3) in a 150-nm thin film as a function of temperature at different positron incident energies. Figure 4 shows results of o-Ps lifetime (or free-volume radius) vs temperature near the surface of 150-nm film. There is a slope change at T_g on the free volume vs temperature plot measured to be 86 ± 2°C which is 11°C lower than the bulk or the center of the thin o-PS films. Figure 5 shows o-Ps vs temperature at nine different depths from near the surface, center, and near the surface of PS film on Au/glass substrate. The intercepts of o-Ps lifetimes are seen to be lowest near the surface, largest in the center of the film, and intermediate low near the interfaces between PS and Au. The resulting T_g variation as a function of the depth is shown in Table 2. The free-volume expansion coefficients ($\beta = \Delta V_f / V_f T$) above and below T_g at different depths of the film, along with the bulk PS data, are listed in Table 2. It is interesting to observe nearly the same β_r in the rubbery state for different depths, but a decreasing β_g in the glassy state as the depth increases from the surface to the center, then increasing again in the interface. It is noted that free-volume expansion coefficients are one order of magnitude larger than specific volume data (on the order of 10^{-4} K^{-1}).

For a thick polystyrene film, T_g was determined to be 97 ± 2 °C by PAL [8,18], which is close to the DSC measurement (100 ± 1 °C). For the thin PS on Au film, we observe a significant suppression of T_g near the surface (86 ± 5 °C) and in the interface (92 ± 2 °C), but no suppression at the center of the film (97 ± 2 °C). The current result of lower T_g near the surface and interface is consistent with the reported T_g suppression near the surface in a thick film [8]. This is consistent with our previous result for an 80-nm PS film on Si substrate [17], which also exhibits a weak interaction with PS. However in an earlier report from another group no T_g variation was observed at different depths [7]. It is very possibly due to the large lifetime resolution from their PAL spectrometer (on the order of >550 ps) than ours, 250-300 ps for PS on Si [17] and 375-500 ps in current studies for PS on Au, and T_g variation escaped their detection.

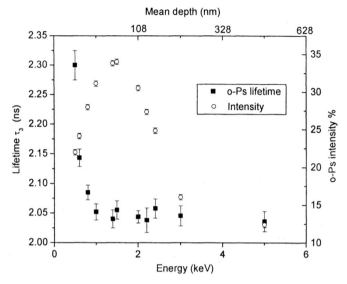

Figure 3.. Variations of o-Ps lifetime (left y-axis) and o-Ps intensity (right y- axis vs the positron incident energy (depth) for a 150 nm PS film on gold coated glass substrate.

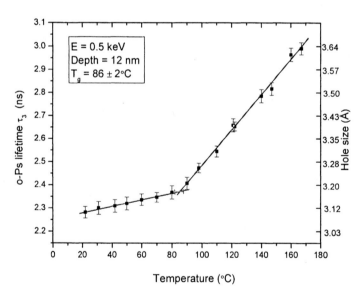

Figure 4. Variation of o-Ps lifetime(left y-axis) or free-volume hole radius (right y-axis) vs temperature at 12 nm from the surface in a 150 nm PS film on gold coated glass substrate. The lines are from linear fitting the data in two regions of temperature.

TABLE 2. T_g, free volume thermal expansion coefficients in a 150-nm polystyrene film on a gold-coated slide.

Locations	T_g (PAL)	$\Delta V_f/V_f T$ (below T_g), K^{-1}	$\Delta V_f/V_f T$ (above T_g), K^{-1}
12 nm	86 ± 2 °C	$(1.3 \pm 0.2) \times 10^{-3}$	$(7.6 \pm 0.2) \times 10^{-3}$
16 nm	90 ± 3 °C	$(1.1 \pm 0.2) \times 10^{-3}$	$(7.5 \pm 0.3) \times 10^{-3}$
25 nm	93 ± 2 °C	$(9.0 \pm 0.3) \times 10^{-4}$	$(7.5 \pm 0.3) \times 10^{-3}$
36 nm	97 ± 2 °C	$(9.2 \pm 0.3) \times 10^{-4}$	$(7.2 \pm 0.4) \times 10^{-3}$
62 nm	97 ± 2 °C	$(9.2 \pm 0.3) \times 10^{-4}$	$(7.5 \pm 0.3) \times 10^{-3}$
70 nm	96 ± 2 °C	$(9.1 \pm 0.2) \times 10^{-4}$	$(7.5 \pm 0.2) \times 10^{-3}$
110 nm	96 ± 2 °C	$(9.1 \pm 0.3) \times 10^{-4}$	$(7.5 \pm 0.4) \times 10^{-3}$
128 nm	94 ± 2 °C	$(1.0 \pm 0.3) \times 10^{-3}$	$(7.6 \pm 0.3) \times 10^{-3}$
148 nm	92 ± 2 °C	$(1.2 \pm 0.2) \times 10^{-3}$	$(7.6 \pm 0.3) \times 10^{-3}$
Bulk of thick film	97 ± 2°C	$(2.0 \pm 0.2) \times 10^{-3}$	$(6.2 \pm 0.3) \times 10^{-3}$

We further analyzed all PAL data as a function of temperature into lifetime distribution for different depths of 150-nm PS film. In Figure 6 we plotted the result of o-Ps lifetime (free-volume radius) distributions at different temperatures at the center (70 nm depth) of the PS film. As expected, the free-volume radius was more widely distributed as the temperature increases due to free-volume expansion. The measured FWHM of the free-volume distributions are plotted vs temperature in Figure 7. It is interesting to observe that the FWHM of the free-volume distribution has an onset temperature similar to that of the free-volume radius plots (Figures 4,5). The onset temperatures from FWHM-temperature plots are close to T_gs as determined from the o-Ps lifetime plot. Figure 8 shows the result of FWHM variations vs temperatures at different depths. A broad free-volume distribution near the surface and at the interface indicates the existence of more end chains and increased polymeric chain mobility as evidenced by the observed lower T_g.

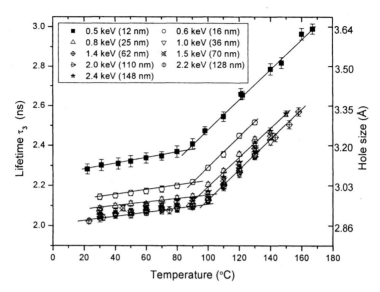

Figure 5. Variation of o-Ps lifetime(left y-axis) or free-volume hole radius (right y-axis) vs temperature at different depth from the surface in a 150 nm PS on gold coated glass substrate. lines are from linear fitting the data in two regions of temperature.

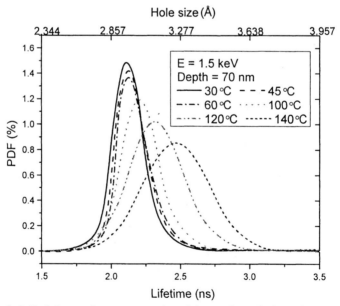

Figure 6. O-Ps lifetime (bottom x-axis) and free-volume hole radius(top x-axis) at different temperature at the center(70nm) of a 150 nm PS on gold coated glass substrate.

The existing reported T_g suppression for a thin film has been interpreted mainly in terms of incomplete entanglement of polymeric chains and broadening of T_g [20,21,45] and by us in terms of free-volume distributions [17]. The new information from the current PAL on another weakly interacting substrate further supports our interpretation that the T_g suppression is caused by free-volume distribution near the surface and in the interface.

The current result of free-volume distribution by PAL offers a new interpretation—that T_g suppression in a thin film is due to different degrees of free-volume distribution at different depths. The free volume has the widest distribution near the surface which leads to large T_g suppression; in the interface, it is slightly broadened but to a lesser extent, which leads to less T_g suppression. In the center of the thin film, it has a distribution similar to that of thick film, which has no observable T_g suppression. The current interpretation is consistent with a theoretical calculation which indicates increased fluctuation of free-volume holes and molecular self-diffusion in nanoscale thin polymer films to account for the T_g suppression [46].

Conclusion

PAS is a novel spectroscopic method, which contains the most fundamental information about physical properties in molecules, i.e. wave function, electron density, and energy level. PAL and DBES techniques provide unique scientific information about the defect properties at the atomic and molecular levels. In this paper we demonstrate the novelty and sensitivity of PAS in measuring the nanoscale-layer structures in thin polymeric films based on obtained free-volume properties. We have observed a significant variation of T_g suppression as a function of depth in a 150-nm polystyrene thin film on Au: 11 K lower near the surface, and 5 K lower in the interface of the Au substrate than the center of the film or in the bulk. This depth-dependence of T_g suppression is further interpreted as a broadening of free-volume distribution in the surface and interfaces.

Acknowledgments

This research has been supported by NSF and NIST. We appreciate Drs. R. Suzuki and T. Ohdaira of AIST, T. N, Nguyen of NIST, and Prof. T.C. Sandreczki for their collaboration.

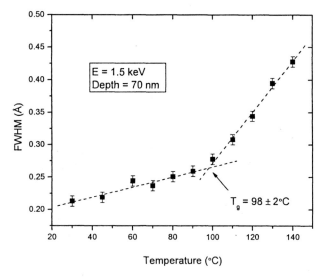

Figure 7. Full-width-at-half-maximum of free-volume hole radius distribution vs temperature at the center of a 150 nm PS on gold coated glass substrate.

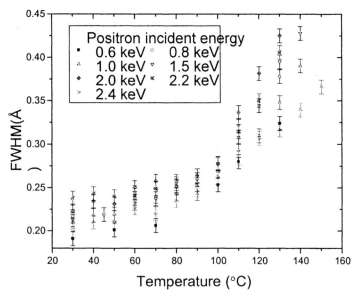

Figure 8. Full-width-at-half-maximum of free-volume hole radius distribution vs temperature at different depths in a 150 nm PS on gold coated glass substrate.

References

1. For example, see *Positron Spectroscopy of Solids*, Dupasquier, A and Mills, Jr., A.P., Eds. (ISO Press, Amsterdam, 1993).
2. For positrons in polymers, see Jean, Y.C. *Microchemical J.* **1990**, 42, 72.
3. For example, see review by Mallon, P.E. in *Principles and Applications of Positron and Positronium Chemistry*, Jean, Y.C.; Mallon, P.E.; Schrader, D.M., Eds., World Sci., Singapore 2003, p. 253.
4. Jean, Y.C.; Dai, G.H.; Suzuki, R. Kobayashi, Y. *AIP Conf. Proc.* **1994**, *303*, 129.
5. Kobayashi, Y.; Kojima, I.; Hishita, S.; Suzuki, T.; Asari, E.; Kitajima, M. *Phys. Rev. B* **1995**, *52*, 823.
6. Xie, L.; DeMaggio, G.B.; Frieze, W.E.; DeVries, J.; Gidley, D.W.; Hristov, H.A.; Yee, A.F. *Phys. Rev. Lett.* **1995**, *74*, 4947.
7. DeMaggio, G.B.; Frieze, W.E.; Gidley, D.W.; Zhu, M.; Hristov, H.; Yee, A.F. *Phys. Rev. Lett.* **1997**, *78*, 1624.
8. Jean, Y.C.; Zhang, R.; Cao, H.; Yuan, J.-P.; Huang, C.-M.; Nielsen, B.; Asoka-Kumar, P. *Phys. Rev. B* **1997**, *56*, R8459.
9. Gidley, D.W.; Frieze, W.E.; Dull, T.L.; Yee, A.F.; Ryan, E. T.; Ho, H.-M. *Phys. Rev. B* **1999**, *60*, R5157.
10. Xie, F.; . Zhang, H.F; . Lee, F.K; Du, B.; Tsui, O.; Yokoe, Y.; Tanaka, K.; Takahara, A.; Kajiyama, T.; He, T. *Macromolecules* **2002**, *35*, 1491.
11. Algers, J.; Sperr, P.; Egger, W.; Kogel, G.; Maurer, F.H.J. *Phys. Rev. B* **2003**, *67*, 125404.
12. Soles, C.L.; Douglas, J.F.; Wu,W.-L.; Peng, H.; Gidley, D.W. *Macromolecules* **2004**, *37*, 2890.
13. Zhang, J.; Zhang, R.; Chen, H.; Li, Y.; Wu, Y.C.; Ohdaira, T.; Sandreczki, T.C.; Suzuki, R.; Jean, Y.C. *Radiat. Phys. Chem.* **2003**, *68*, 535.
14. Zhang, J.; Chen, H.; Zhang, R.; Li, Ying; Suzuki, R.; Ohdaira, T.; Jean, Y. C. *Materials Science Forum* **2004**, *445-446*, 367.
15. Zhang, R.; Cao, H.; Chen, H.M.; Mallon, P.; Sandreczki, T.C.,; Richardson, J.R.; Jean, Y.C.; Nielsen, B.; Suzuki, R,; Ohdaira, T. Radiation Phys. and Chem. **2000**, *58(5-6)*, 639
16. Chen, H.; Zhang, R.; Li, Y.; Zhang, J.; Wu, Y.C.; R.; Sandreczki, T.C.; Mallon, P.E.; Suzuki, R.; Ohdaira, T.; Gu, X.; Nguyen, T.; Jean, Y.C. *Materials Science Forum* **2004**, *445-446*, 274.
17. Jean, Y.C; Zhang, J.; Chen, H.; Li, Y.; Liu, G. *Spectrochimica Acta Part A* **2005**, *61*, 1683.
18. Zhang, Junjie, Ph.D. dissertation from University of Missouri-Kansas City, May, 2005.
19. de Gennes, P.G. *Euro. Phys. J.* **2000**, 2, 201.

104

20. Keddie, J. L.; Jones, R.A.L.; Corey, R. A. *Europhys. Lett.* **1994**, *27*, 59.
21. Forrest, J.A.; Dalnoki-Veress, K.; and Dutcher, J.R. *Phys. Rev. E* **1998**, *58*, 6109.
22. Kawana, S; and Jones, R.A.L. *Phys, Rev. E* **2001**, *63*, 021501.
23. Wallace, W. E.; Van Zanten, J. H.; Wu, W. L. *Phys. Rev. E* **1995**, *52*, R3329.
24. Jones, R.L.; Kumar,S.K.; Ho, D.L.; Briber, R.M.; and Russell, T.P. *Macromolecules* **2001**, *34*, 559.
25. Efremov, M. Y.; Olson, E. A.; Zhang, M. Z.; Allen, L. H. *Phys. Rev. Lett.* **2003**, *91*, 85703.
26. Fryer, D.S.; Peters, R.D.; Kim, E.J.; Tomaszewski, J.E.; de Pabio, J.J.; Nealey, P.F. *Macromolecules* **2001**, *34*, 5627.
27. Bliznyuk, N. V.; Assender, H. E.; Briggs, G.A.D. *Macromolecules* **2002**, *35*, 6613.
28. Fisher, H. *Macromolecules* **2002**, *35*, 3592.
29. Zhang, C.; Hong, S.-C.; Ji, N.; Wang, Y.-P.; Wei, K.-W.; Shen, Y.R. *Macromolecules* **2003**, *36*, 3303.
30. Ellison, C. J.; Torkelson, J. M. *Nature Mat.* **2003**, *2*, 695.
31. White, C.C.; Migler, K.B.; ;Wu, W.L.; *Poly. Eng. and Science* **2001**, *41*, 1497.
32. PATFIT package (1989) purchased from Riso National Laboratory, Denmark.
33. Cao, H.; Zhang, R.; Zhang, J.-P.; Huang, C.-M.; Jean, Y.C.; Suzuki, R.; Ohdaira, T.; Nielsen, B. *J. Phys. C: Condens. Matt.* **1998**, *10*, 10429.
34. Zhang,R.; Gu, X.; Chen, H.; Zhang, J.; Li, Y.; Nguyen, T.; Sandreczki, T.C.; Jean, Y.C. *J. Polym. Sci. B: Polym. Phys.* **2004**, *42*, 2441.
35. Jean, Y.C. *Macromolecules* **1996**, *27*, 5756.
36. Tao, S. J. *J. Chem. Phys.* **1972**, *56*, 5499.
37. Nakanishi, N.; Wang,, S.J.;Jean, Y.C. in *Positron Annihilation Studies of Fluids*, Sharma, S.C., Ed., World Scientific, Singapore 1988, p. 292.
38. Ito, K.; Nakanishi, H.; Ujihira, Y. *J. Phys. Chem B*, **1999**, *103*, 4555.
39. Dull, T. L.; Frieze, W. E.; Gidley, D. W.; Sun, J. N.; Yee, A. F. *J. Phys. Chem. B*, **2001**, *105*, 4657.
40. Cao, H.; Dai, G. H.; Yuan, J.-P.; Jean, Y.C. *Mater. Sci. Forum* **1997**, *238*, 255.
41. Kansy, J. *Nucl. Instr. Meth. A* **1996**, *374*, 235.
42. Shukla, A.; Peter, M.; Hoffman, L. *Nucl. Inst. Meth. A* **1993**, *335*, 310.
43. Schultz, P.J.; Lynn, K.G. *Rev. Mod. Phys.* **1998**, 60, 701.
44. Van Veen, A.; Schut, H.H.; de Vries, J.; Hakvoort, H.A.; IJpma, M.R. *AIP Conf. Proc.* **218**, 171 (AIP, New York, 1990).
45. Forrest, J.A.; Dalnoki-Veress, K.; Dutcher, J.R. *Phys. Rev. E*, **1997**, 56, 5705.
46. Chow, T.S. *J. Phys. C: Condens. Matt.* **2002**, 14, L333.

Chapter 8

Recent Advances in Film Formation from Colloidal Particles: Synthesis of Non-Spherical Colloidal Shapes with Stimuli-Responsive Characteristics

Marek W. Urban

School of Polymers and High Performance Materials, The University of Southern Mississippi, Hattiesburg, MS 39406

Abstract

The synthesis of colloidal dispersions using biologically active dispersing agents that exhibit stimuli-responsive characteristics lead to unique particle shapes and surface/interfacial morphologies developed during coalescence. This process enables the reparation of a wide range of particle shapes of unique composition and architecture. By tuning parameters such as concentration levels and chemical structure of phospholipids tailored properties could be achieved. Recent advances in film formation from non-spherical colloidal particles are discussed in the context of controlled stratification stimulated by thermal, ionic strength, and pH stimuli.

Introduction

A vast majority of the recognition and response driven biological processes are initiated at or near surfaces or interfaces, and one of the examples are molecular engines generating mechanical energy powered by chemical processes that result from the swelling and shrinkage of macromolecular segments, thus causing motion. Spasmonenes are a type such engines in which the imbalance between osmotic and entropic forces [1] resulting from stored energy between electrostatic repulsions and negatively charged filaments facilitates the motion.[2] When Ca^{2+} ions are present, the highly charged state of spasmonenes is neutralized and the filaments collapse entropically to form a rubber-like material. Similar stimuli-responsive behavior was recently recognized in colloidal dispersions stabilized by phospholipids,[3] which upon coalescence can be driven to either film-air (F-A) or film-substrate (F-S) interfaces, and the primary stimuli are pH, ionic strength of the colloidal solution, and/or enzyme concentration levels. The focus on (co)polymer systems that are responsive to external stimuli are of particular interest because these entities, if properly designed, may become responsive to stimuli similar to biological systems, thus creating scientific and technological opportunities in fields as diverse as water treatment, enhanced oil recovery, controlled drug release, biomedical applications, personal care products, and home-land security, to name just a few.

Polymeric films and coatings offer numerous applications ranging from commodity materials to highly sophisticated self-healing and/or self-releasing films. To produce traditional colloidal dispersions monomers and surfactants are utilized and synthetic efforts will determine particle morphologies ranging from randomly polymerized copolymers to core-shell and other morphologies. As shown in Figure 1, colloidal dispersions can be prepared by simple synthetic efforts, and synthetic conditions may lead to core-shell morphologies.[4] Although these synthetic efforts may appear rather trivial, it is the design of colloidal particles that determines not only solution properties of colloidal dispersions, but also their coalescence.

In an effort to form polymeric films, colloidal particles, upon water evaporation, and providing that the glass transition temperature of the particles is below the minimum film formation temperature (MFFT), must coalesce. Primarily curiosity driven, numerous studies have been conducted on film formation which led to more of less to similar conclusions: coalescence is a film formation process influenced by many internal and external factors, and the classical models of dry sintering,[5] capillary theory,[6] wet sintering,[7] or particle deformation[8] adequately describe a physical phenomenon. Consequently, a number of reviews on latex film formation flooded the literature which reduced this process to three drying stages of latex.[9]

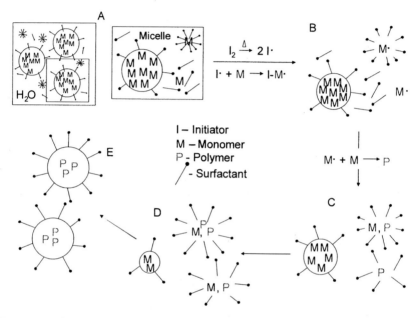

Figure 1. Schematic diagram illustrating the synthesis of traditional spherical colloidal particles, where in the presence of water, initiator (I), and surfactant, monomer (M) is polymerized to form water-dispersible colloidal particles. (See page 1 of color inserts in this chapter)

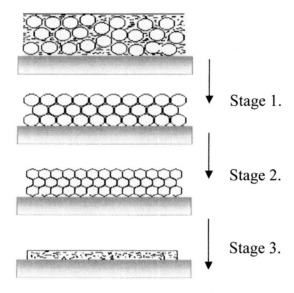

Stage 1.

Stage 2.

Stage 3.

Figure 2. Simplified stages of coalescence of colloidal particles.
(See page 2 of color inserts in this chapter)

Figure 2 illustrates a summary of these efforts which, as was proposed, consists of three stages: (1) water evaporation during which particle concentration increases; (2) particles come into contact with each other and become deformed, and (3) particles diffuse into each other to form entanglements. This oversimplified description of the film formation neglects complexity of chemistries and interfacial phenomena involved in colloidal solutions from which films are formed and consequently, stratification processes resulting from mass transport during coalescence. This chapter focuses on recent advances in colloidal particle design giving raise to unique solution morphologies as well as stimuli-responsive coalescence of coatings.

Particle Morphologies and Film Formation

Although it is often assumed that polymeric films are uniform, over a decade ago it was shown that compatibility of individual components, glass, transition temperature, coalescence conditions, and other variables, play a key role in film formation.[10,11] Consequently, surface and interfacial properties are influenced. In an effort to generate desirable low surface tension films from colloidal particles, the use of fluorine-containing surfactants along with synthetic efforts have led to unique particle morphologies which are shown in Figure 3.

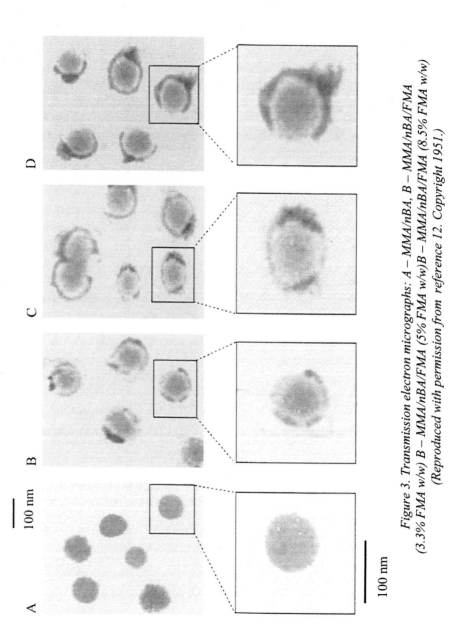

Figure 3. Transmission electron micrographs: A – MMA/nBA, B – MMA/nBA/FMA (3.3% FMA w/w) B – MMA/nBA/FMA (5% FMA w/w) B – MMA/nBA/FMA (8.5% FMA w/w) (Reproduced with permission from reference 12. Copyright 1951.)

100 nm

100 nm

In this particular study,[12] a simple synthetic method for preparing stable colloidal dispersions that form polymeric films containing fluorinated-acrylates was developed, in which the simultaneous presence of the dual tail anionic fluorosurfactant phosphoric acid bis(tridecafluoro-octyl) ester ammonium salt (FSP) and sodium dodecyl sulfate (SDS) surfactants facilitates a suitable environment for the aqueous polymerization of methyl methacrylate/n-butyl acrylate/heptadeca fluoro decyl methacrylate (MMA/nBA/FMA) colloidal particles. Polymerization was achieved by a low-sheer monomer-starved emulsion polymerization process in which the FSP/SDS surfactant mixture significantly reduced the surface tension of the aqueous phase, thus facilitating mobility and subsequent polymerization of FMA along with MMA and nBA momomers. Using this approach, it is possible to obtain stable non-spherical MMA/nBA/FMA colloidal dispersions containing up to 8.5% (w/w) copolymer content of FMA. Furthermore, surface properties with a significant decrease in the kinetic coefficient of friction as well as high contact angles can be produced. As shown in Figure 4, a significant drop of the kinetic coefficient of friction is achieved as the concentration of FMA increases. At the same time, MEK resistance is enhanced. More recent studies showed that the use of phospholipids as dispersing agents facilitates environments to further increase FMA in colloidal particles and can be as high as 15 w/w%.[13] Such dispersions after film formation exhibit very low kinetic coefficient of friction which is accomplished by vertical phase separation between pMMA/nBA and p-FMA phases, thus creating "Teflon™-like" surface properties. It should be noted that these dispersions are extremely stable.

Particle Shapes and Film Formation

Recent conceptual and experimental advances in colloidal particle design which take advantage of the molecular structures of dispersing agents and preparation techniques of colloidal particles have led to the development of unique particle morphologies and films with stimuli-responsive characteristics. As indicated above, the use of phospholipids in the synthesis of colloidal particles was only recently discovered, which created a number of opportunities for the development of unique colloidal shapes. Figure 5 schematically illustrates the difference between the particle shape obtained using traditional surfactants and biologically active phospholipids. As seen, majority of surfactants form micelles, whereas phospholipids, due to their unique hydrophobic-hydrophilic interactions may form liposomes or tubules, and the nature of these organized features depends on their chemical makeup.

It is possible to create hollow particles when phospholipids are utilized in the synthesis of reactive monomers and the degree of the wall reinforcement will depend upon the degree of polymerization. As shown in Figure 6, hollow particles can be prepared using this approach and their size may be also altered

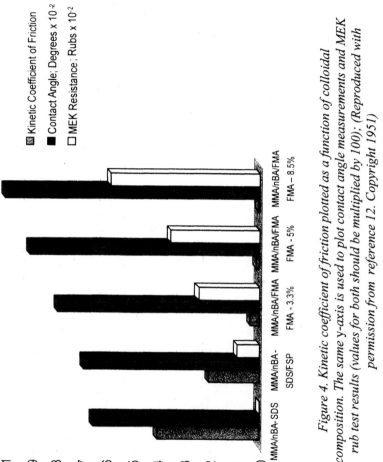

Figure 4. Kinetic coefficient of friction plotted as a function of colloidal composition. The same y-axis is used to plot contact angle measurements and MEK rub test results (values for both should be multiplied by 100). (Reproduced with permission from reference 12. Copyright 1951)

112

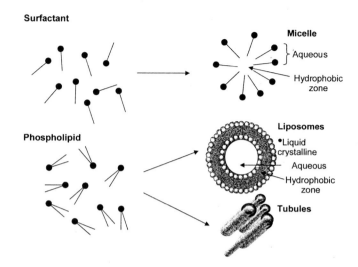

Surfactant

Micelle

Aqueous

Hydrophobic zone

Phospholipid

Liposomes

•Liquid crystalline

Aqueous

Hydrophobic zone

Tubules

*Figure 5. Formation of micelles, liposones, or tubules that may serve as templates for emulsion polymerization.
(See page 2 of color inserts in this chapter)*

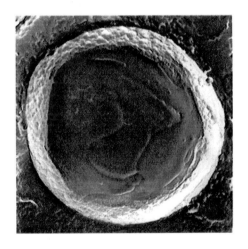

500 nm

Figure 6. Hollow colloidal particle of polystyrene obtained in the presence of a liposome from 1,2-dilauroyl-phosphocholine (DLPC) phospholipid in an aqueous phase (Reproduced with permission from reference 14. Copyright 2005 American Chemical Society.)

using a nano-extrusion process. [14] In order to create tubular morphologies a number of studies have been conducted, in particular, on polymerizable phospholipids such as 1,2-bis(10,12-tricosadiynoyl)-*sn*-glycero-3-phosphocholine (DC$_{8,9}$PC). These species may be polymerized in an aqueous phase upon exposure to UV radiation, and although mechanistic aspects leading to their formation are still debatable, as illustrated in micrographs shown in Figure 7, A-D, an exposure for 5, 10, 60, and 120 min to 254 nm UV radiation generates shapes that change from random tubules (A) to twisted (B,C) and helical (D) shapes. The color of the lipid solution also gradually changes from milky (A-unpolymerized), to light reddish (5-10 min) and intense red (60-120 min). This is also illustrated in Figure 7, B, and these results indicate that random, initially dispersed tubules have polymerized as a result of the UV exposure through the reactions of diacetylenic groups to form conjugated C=C networks.

While these recent examples demonstrate that the role of dispersing agents not only determines solution properties, but they also have a significant effect on shapes, size, and morphology of particles. Although obtaining other shapes represents another level of challenges, recent studies reported the synthesis leading to *cocklebur-shape* particles which may open new avenues for combining man-made polymers with biological systems. As shown in Figure 8, SEM images exhibit a cocklebur-shape that consists of the spherical core and tubular-shape rods sticking outwards. To obtain these unique shapes a nano-extruder combined with specific phospholipid chemistries incorporated in the synthesis process was utilized. [15] Although the use of the extruder is not necessary in controlling particle morphologies, in this case its use allowed the passage of the dispersed solution from one reservoir to another through precisely sized membrane channels.

As a result, mono-dispersed droplets were obtained, which upon polymerization, resulted in colloidal particles. While the size of the extruder channels may be varied from nano to micro range, thus dictating the core size of the particles, the choice of phospholipids determines the unique cocklebur morphologies. Cocklebur morphologies can be obtained by utilizing 1,2-bis(10,12-tricosadiynoyl)-*sn*-glycero-3-phosphocholine (DCPC) phospholipids in the presence of azo-bis-isobutyronitrile (AIBN) initiator dissolved in methyl methacrylate (MMA) and n-butyl acrylate (nBA) monomers, and sodium dioctyl sulfosuccinate (SDOSS) surfactant. The same process can produce sizes ranging from a few nanometers up to several microns. The unique feature of the utilization of DCPC are temperature responses. Below 38°C, DCPS itself forms crystalline tubules, which has been demonstrated in the literature. [16,17,18] When heated above 38°C, DCPC molecules undergo a transition from a crystalline to a liquid crystalline phase. Thus, heating the pre-extruded droplets of monomers stabilized by DCPC and SDOSS to 75°C facilitates polymerization of MMA and nBA monomers, but subsequent cooling generates the cocklebur particle architectures exhibiting cocklebur-shape morphologies with photopolymerizable DCPC crystalline tubules projecting radially from the particle surface. The use

114

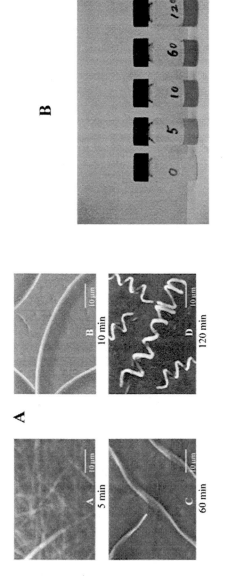

Figure 7. A. Scanning electron micrographs of DC$_{8,9}$PC exposed for 5, 10, 60, and 120 min to 254 nm UV radiation. B. optical images of polymerized phospholipids DC$_{8,9}$PC placed in an aqueous phase after UV exposure for 0, 5, 10, 60, and 120 min. (See page 3 of color inserts in this chapter)

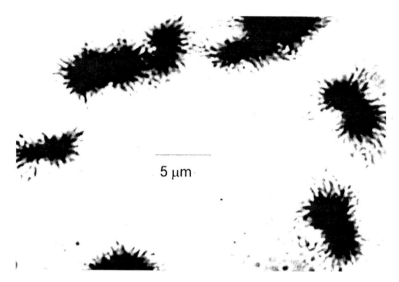

5 μm

*Figure 8. A - TEM micrograph at 3,000X mag. of large MMA/nBA copolymer
particles stabilized by SDOSS/DCPC mixture; B - TEM micrograph at 12,000X
mag. of large MMA/nBA copolymer particles exhibiting DCPC phospholipid
tubules at the interface (Reproduced from reference 15. Copyright 2006
American Chemical Society.)*

of classical emulsion polymerization processes lead to similar particle
morphologies, thus cocklebur shapes are not attributed to the pre-extrusion
process, but the simultaneous presence of SDOSS/DCPC during polymerization
of monomers and these morphologies are a function of ionic strength of the
solution. The same particles stabilized with a SDOSS/DCPC mixture containing
$CaCl_2$ (0.018 %w/w) do not exhibit cocklebur surfaces. Considering the fact that
surface tubules are sensitive to temperature changes, particularly around 38°C,
and the formation of cocklebur morphologies is responsive to the ionic strength
of an aqueous phase. These shapes provide numerous biological and biomedical
opportunities, as one can envision placing various biological sensors inside or
outside the tubules capable of recognizing targeted bio-systems, whereas the
interior part of the particle may contain drugs or other species.

In view of the above considerations it should be realized that traditional
concepts of coalescence processes do not take into account different shapes and
more complex surface energies, thus film formation processes cannot be reduced
to simple particle interdiffusion. In the case of the cocklebur particles
permeability experiments conducted as a function of UV exposure were
performed in which aqueous p-MMA/nBA films containing DCPC deposited
and irradiated with 254 nm UV radiation at 0-150 min exposure times. As
shown in Figure 9, Curves A and B, both p-MMA/nBA copolymer and p-

MMA/nBA copolymer containing 0.3% DCPC films not exposed to UV (0 min) exhibit 70 wt.% loss, which is not surprising as all colloidal dispersions contain 32 w/w % solids. However, as UV exposure time increases, copolymer films containing DCPC (Curve B) undergo decreased wt.% mass loss, and at 150 min UV exposure and 72 hrs coalescence time, these films exhibit 54 wt.% total mass loss compared to 70 wt.% total mass loss by the copolymer (Curve A). These data illustrates that the water retention can be fairly substantial due to crosslinking reactions initiated by UV radiation.

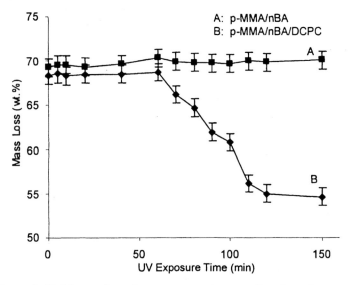

Figure 9. Wt.% mass loss due to evaporation as a function of exposure to UV radiation for times of 0-150 min in (A) p-MMA/nBA copolymer and (B) p-MMA/nBA copolymer containing 0.3% DCPC films.

The use of phosholipids as co-surfactant in colloidal dispersion synthesis presents the opportunity for formation of multiple micellar environments which serve as polymerization loci to produce uni-, bi-, or even multi-modal particle sizes, depending upon the number and size of micelles generated. If both surface stabilizing species are immiscible within each other, two sizes of micelles can be obtained.[19] In contrast, for SDOSS/MHPC (1-myristoyl-2-hydroxy-sn-glycero phosphocholine) uni-modal colloidal particles are obtained and, when such colloidal dispersions are allowed to coalesce under various stimuli conditions, unique surface localized ionic clusters are obtained.

Figure 10 illustrates atomic force microscopy (AFM) and internal reflection infrared (IRIR)[20] images recorded from the F-A interface of SDOSS/MHPC stabilized p-MMA/nBA films annealed at 150°C. As shown in the AFM images,

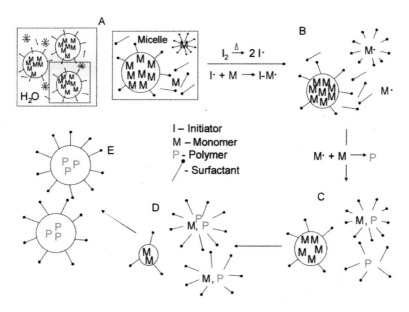

Figure 8.1. Schematic diagram illustrating the synthesis of traditional spherical colloidal particles, where in the presence of water, initiator (I), and surfactant, monomer (M) is polymerized to form water-dispersible colloidal particles.

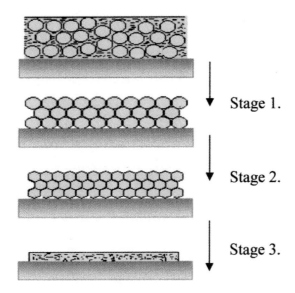

Figure 8.2. Simplified stages of coalescence of colloidal particles.

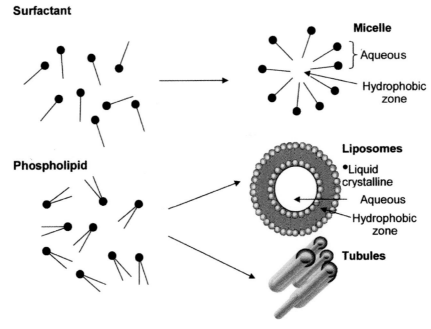

Figure 8.5. Formation of micelles, liposones, or tubules that may serve as templates for emulsion polymerization.

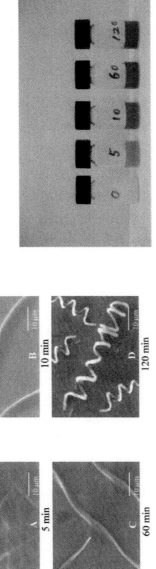

Figure 8.7. A. Scanning electron micrographs of DC$_{8,9}$PC exposed for 5, 10, 60, and 120 min to 254 nm UV radiation. B. optical images of polymerized phospholipids DC$_{8,9}$PC placed in an aqueous phase after UV exposure for 0, 5, 10, 60, and 120 min.

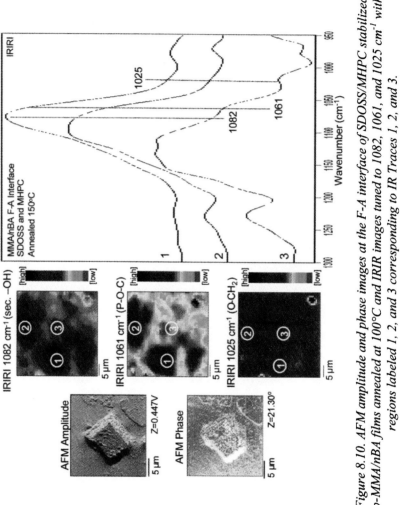

Figure 8.10. AFM amplitude and phase images at the F-A interface of SDOSS/MHPC stabilized p-MMA/nBA films annealed at 100°C and IRIR images tuned to 1082, 1061, and 1025 cm⁻¹ with regions labeled 1, 2, and 3 corresponding to IR Traces 1, 2, and 3.

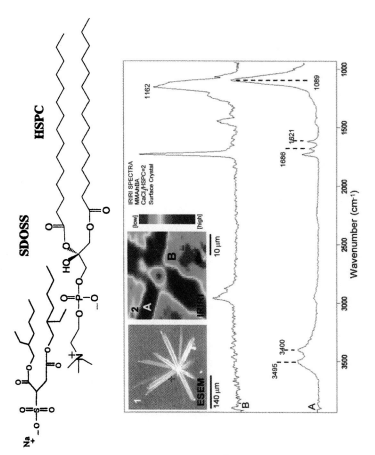

Figure 8.11. ESEM micrograph (Image 1) and IRIRI spectra (Image 2) of surface crystals at the FA interface of coalesced film cast from MMA/nBA colloidal dispersion treated with 2.0/1.0CaCl₂/HSPC ratio: (A) IRIRI contact on the surface crystal; (B) IRIRI contact on film surrounding the surface crystal

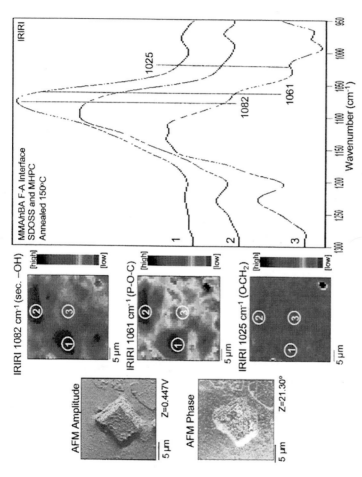

Figure 10. AFM amplitude and phase images at the F-A interface of SDOSS/MHPC stabilized p-MMA/nBA films annealed at 100° C and IRIR images tuned to 1082, 1061, and 1025 cm^{-1} with regions labeled 1, 2, and 3 corresponding to IR Traces 1, 2, and 3. (See page 4 of color inserts in this chapter.)

MHPC

SDOSS

Structures for MHPC and SDOSS.

temperature effectively stimulates the formation of surface-localized ionic clusters (SLICs) 21 which exhibit ordered crystalline geometries with 5 x 10 um dimensions at the F-A interface. Areas labeled 1-3 in the IRIR images, from which multiple images were collected, illustrate that when tuned to 1082 (-OH) and 1061 (P-O-C) cm^{-1}, IRIR images exhibit the structures observed with AFM. These data indicate that SLICs generated at the F-A interface with annealing at 150°C contain higher concentration levels of the tuned entities, thus revealing that the primary component of these surface structures is MHPC. Interestingly enough, spectral Traces 1 and 2, corresponding to Regions 1 and 2, exhibit elevated intensities at 1082 and 1061 cm^{-1} which indicate the presence of –OH and P-O-C segments of MHPC. The spectral map generated when tuned to 1025 cm^{-1} due to O-CH2 band in SDOSS exhibits the same SLIC heterogeneities. However, the concentration of O-CH2 present in areas 1-3 is lower, thus revealing a minor contribution of SDOSS to SLIC formation at 150°C.

The situation changes during coalescence when dual tail phospholipids are utilized.[22] For example, the presence of hydrogenated soybean phosphatidylcholine (HSPC) and sodium dioctyl sulfosuccinate (SDOSS) (shown below) has significant effect on mobility of individual components during coalescence. The presence of HSPC inhibits migration of SDOSS to the F-A interface, but the presence of electrolyte species such as aqueous CaCl$_2$, has a very pronounced effect on film formation. For example, when the Ca^{2+}/HSPC ratio is 0.1/1.0, SDOSS is released to the F-A interface during coalescence, but at 2.0/1.0 Ca^{2+}/HSPC, HSPC diffuses to the F-A interface and crystalline domains consisting of HSPC are formed. Figure 11, Image 1, illustrate scanning electron microscopy (SEM) image recorded from SLICs produced at the F-A interface during coalescence. To identify chemical features of crystalline domains IRIR images were collected and are depicted in Figure 11, Image 2. The images were obtained from the same area from which the SEM image was detected. Spectroscopic analysis of areas A and B of Figure 11, Image 2, resulted in the spectra labeled A and B of Figure 11. Trace A is due to HSPC represented by the band at 1089 cm^{-1} due to P-O segments, whereas Trace B is due to the surrounding MMA/nBA copolymer matrix with the band at 1162 cm^{-1} due to O-C-C stretching vibrations. These observations clearly indicate that the crystals formed when the Ca^{2+}/HSPC ratio is equal to 2.0/1.0 results in the release of HSPC to the surface and its crystallization that occurs during coalescence.

This stimuli-responsive behavior illustrates that it is possible to control the release of different surface active species during coalescence which in this case results in the formation of SLIC domains. However, other components may also stratify and numerous previous studies outlined chemical and physical factors responsible for this process.[23]

In summary, the choice of components during colloidal synthesis determines not only morphologies of colloidal particles, but their responsiveness during and after coalescence. When particles coalesce to form polymeric films, they are capable of forming locally organized interfacial structures which are affected not only by their compatibility of the components, but also by electrolyte environments of colloidal dispersions, temperature, enzyme concentrations and activity, protein adsorption, and other physical stimuli. In this chapter the main focus was on Na^+, K^+, and Ca^{2+} counter ions which exhibit significant influence on coalesced films, where different degrees of diffusion of phospholipids to the interfaces and their conformational changes may occur, depending upon ionic strength of the solution. For example, in contrast to other ions, the presence of Ca^{2+} results in organized structures that collapse entropically to form random surface layers, similar to those produced in nature. Regardless of the driving forces for mobility of individual components in polymeric films, spatial compositional and structural gradients exhibit considerable interest in designing biointerfaces, tribology, biomedicine, biochemical processes, drug delivery, and others.

Acknowledgments

These studies were partially supported by several industrial sponsors and primarily by the MRSEC Program of the National Science Foundation under Award Number DMR 0213883. The Instrumentation Program of the National Science Foundation under the Award Number DMR 0215873 is also acknowledged for partial support of these studies.

SDOSS

HSPC

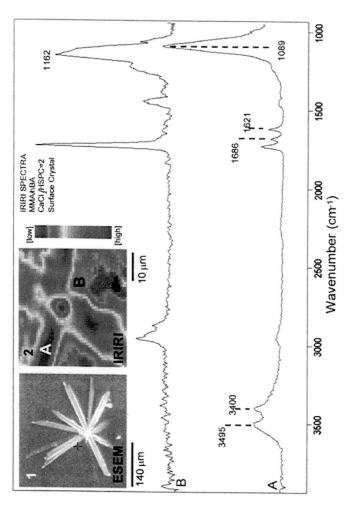

Figure 11. ESEM micrograph (Image 1) and IRIRI spectra (Image 2) of surface crystals at the FA interface of coalesced film cast from MMA/nBA colloidal dispersion treated with 2.0/1.0CaCl₂/HSPC ratio: (A) IRIRI contact on the surface crystal; (B) IRIRI contact on film surrounding the surface crystal (See page 5 of color inserts in this chapter.)

122

References

1. Moriyama, Y., Okamoto, H., Asai, H., *Biophys. J.*, **1999**, 76, 993.
2. Hofmann-Berling, H., *Biochim.Biophys., Acta,* **1958**, 27, 247.
3. Jacoubs, A. Urban, M.W. *Biomacromolecules*, **2003**, 4, 52-56.
4. Zhao, Y.; Urban, M.W. *Langmuir* **2000**, 16 (24), 9439-9447.
5. Dillon, R.E.; Matheson, L.A.; Bradford, E.B. *J. Colloid Sci.* **1951**, 6, 108.
6. Brown, G.L. *J. Polym. Sci.* **1956**, 22, 423.
7. Vanderhoff, J.W.; Tarkowski, H.L.; Jenkins, M.C.; Bradford, E.B. *J. Macromol. Chem.* **1966**, 1, 131.
8. Sheets, D.P. *J. Appl. Polym. Sci.*, **1965**, 9, 3759.
9. Holl, Y.; Keddie, J.L.; McDonald, P.J.; Winnik, M. in *Film Formation on Coatings*, ACS Symp. Series #790, pp.2-26. Eds. Provder, T.; Urban, M.W. Amer. Chem. Soc., Washington DC, **2001**.
10. Evanson, K.; Urban, M.W., *Polym. Comm.*, **1990**, 31, 279.
11. Zhao, Y.; Urban, M.W., *Film Formation in Coatings*, ACS Symp. Series #790, Eds. Provder, T.; Urban, M.W., Amer. Chem. Soc., **2001**, Washington, DC.; and ref. therein.
12. Dreher, R.W., Singh, A.; Urban, M.W., *Macromolecules*, **2005**, 38, 4666.
13. Singh, A.; Dreher, R.W., Urban, M.W., *Biomacromolecules*, **2005**, submitted.
14. Lestage D.J.; Urban, M.W., *Langmuir*, **2005,** 21. 4266.
15. Lestage D.J.; Urban, M.W., *Langmuir*, **2005**, in press.
16. Yager, P.; Schoen, P. E. *Mol. Cryst. Liq. Cryst.* **1984**, *106*, 371-381.
17. Svenson, S.; Messersmith, P. B. *Langmuir* **1999**, *15*, 4464-4471.
18. Thomas, B. N.; Safinya, C. R.; Plano, R. J.; Clark, N. A. *Science* **1995**, *267*, 1635.
19. Lestage, D. J.; Urban, M. W. *Langmuir* **2004**, *20*, 7027-7035.
20. Otts, D.; Zhang, P.; Urban, M.W., *Langmuir*, **2002**, 18, 6473.
21. Dreher, R.W.; Urban, M.W., *Macromolecules*, **2005**, 38, 2205.
22. Lestage, D.J.; Urban, M.W. *Langmuir,* **2004**, 20, 7027.
23. Urban, M.W. *Stimuli-Responsive Polymeric Films and Coatings*, ACS Symp. Sereis #912, Ed. M.W. Urban, Amer. Chem. Soc., Washington, DC, **2005**, and ref. therein.

Chapter 9

Formation of Films of Two-Dimensional Continuous Network Skeleton

Xiaorong Wang[1] and Naruhiko Mashita[2]

[1]Center for Research and Technology, Bridgestone Americas,
1200 Firestone Parkway, Akron, OH 44317–0001
(wangxiaorong@bfusa.com)
[2]Chemical and Industrial Products, Materials Development Department,
Bridgestone Corporation, 1 Kashio-cho, Totsuka-ku, Yokohama,
Kanagawa 244–8510, Japan

A noval technology has been developed which creates thin films that consist of essentially two-dimensional polymeric skeleton-like reticulated networks of cell sizes of several micrometers. The formation of the two-dimensional network was initiated by the spinodal decomposition of a binary polymer/solvent mixture in thin films, and the developed phase pattern was then preserved by the crystallization of the polymer in polymer-rich phases. The key principle in the technology was the fine-tuning of the solvent quality for that the crystallization of the polymer occurs just below the miscibility gap. Varying the cooling rate, solvent quality, film thickness and the composition can control the cell size.

Spinodal decomposition (SD) in polymer solutions produces fascinating network structures or patterns (1-15). The reason is that the polymer-rich phase (or the dynamically slow component) during phase separation becomes more and more viscoelastic with time. The initial growth of the concentration fluctuations in those systems is soon suppressed by the formation of the slow-component phase (1-9). The resultant process is then governed by a competition between thermodynamics and visco-elastics. In others words, the formation of the final morphological structure is usually a consequence of a unique combination of thermodynamics and visco-elastics. Of importance is that the pattern evolutions observed in a polymer solution are essentially the same to that in a polymer blend of different T_gs (4). Understanding the phenomenon and the process is pertinent because they are of intrinsic importance not only in industrial materials developments (1, 9-10) but also in fundamental investigations (16-18). There has been considerable theoretical and experimental work on describing the characteristic features of those systems (18-23). In this study, we present a novel experimental approach to form films of two-dimensional networks through the phase separation of a polymer solution in thin slabs of thickness of micrometers. We showed that by suitably selecting the thickness, cooling procedure and solvent quality, we obtained films of essential two-dimensional (2D) continuous network skeleton of the pure polymer.

Experimental

Semi-crystalline poly(ethylene-co-propylene) rubber (or c-EPR) was from JSR under commercial name EP01. The polymer was made via transition-metal-catalyzed polymerization of ethylene and propylene. The ethylene content in the polymer was about 60%. The average molecular weight of the material Mw= 2.1×10^5 with polydispersity index of 2.15. Here, the melting point T_m of the polymer was 52°C and the crystallinity of the polymer was about 12%. The polymer was used without further purification.

Diisodecyl adipate (or DIDA), 99% purity, was from the C. P. Hall Company. The material was a clear colorless liquid, and had a viscosity of about 30 cps and a density of 0.918 g/cm^3 at 23°C. Prior to the use, the DIDA solvent was filtered through a micro-membrane of about 0.2µm pore size. This precaution is important for minimizing dust effects on phase separation. Other high-boiling point solvents, such as the trioctyl phthalate (TOP), were also pre-treated in the same fashion.

The c-EPR polymer was first swelled and then dissolved into the DIDA solvent at approximately 150°C. Complete dissolution took place within one hour with vigorous stirring. Once it became clear, the solution was then diluted to 12 wt% concentration. The solution was homogeneous and transparent at temperature T>110°C. The phase separation temperature of the solution was 105°C. The crystallization temperature of the polymer in the DIDA solution was about 43°C. Figure 1 shows the phase diagram of the c-EPR/DIDA system. This system was slightly different from that of our previous report (*1*). The merit of using this system is that there is a large gap between the de-mixing temperature and the crystallization temperature of the polymer in the solution. This facilitates one to study the phase evolution without worrying about interferences from the crystallization of the polymer. Phase separation in such a system can be easily initialized by quenching the solution to the unstable part of the diagram (see the procedure displayed in Figure 1), and the phase pattern can be frozen at any moment for detail investigations by crystallization of the polymer in the polymer-rich-phase. Using a solution of 12 wt% polymer simply ensures that the phase separation is SD-induced type.

To make a wedge-like thin film, a small drop of the hot solution was first placed on a pre-cleaned 25mm wide x 75mm long x 1mm thick microscope glass slide. On the top of the liquid drop was then covered with a microscope cover glass of 22 wide x 26mm long x 0.17mm thick. Between the two glasses, a strip-like spacer of thickness of about 10 μm was used on the one side of the microscope slides. The liquid that was sandwiched between the two slides was then pressed to form a wedge-like liquid film. The thickness of the film changed gradually from 0 to 10 μm over a length of 25mm. The preparation was carried on a hot stage of temperature higher than 150°C.

To make a planar thin film, a small drop of the hot solution was first placed on a pre-cleaned 10mm diameter x 0.2mm thick microscope glass that was glued on a spin-coater SCX-50 (from Novocontrol GmbH). Then, the planar thin film was made by letting the small drop to spin at a speed ranging from 1000 to 6000 rpm. A heat gun was used to keep the surface temperature of the glass higher than 150°C. An IR laser thermometer (from Extech) was used to monitor the surface temperature. Variation in film thickness was achieved by varying the temperature and the spin speed.

Phase separation in those films was monitored using a Carl Zeiss optical microscope equipped with a Mettler FP52 hot stage that was controlled by a Mettler FP90 control processor. Usually a sample film was first equilibrated at 150°C for 20 minutes and was then quenched to 60°C. Phase separation was developed after the solution was held at 60°C for 10 minutes. The developed phase pattern was then preserved by cooling down to 23°C at a cooling rate 10°C/min. Phase picture in the film was imaged at 23°C using a instant Polaroid camera that had been loaded with plate-like sheet films.

Results and Discussion

The c-EPR/DIDA solution is homogeneous and transparent at temperature T>110°C. It can turn into opaque once it is brought to a temperature T≤100°C. Initially, usual growth of concentration fluctuations occurs throughout the

126

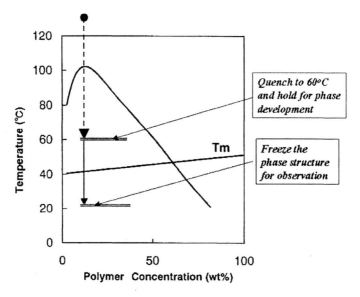

Figure 1. Phase diagram for semi-crystalline poly(ethylene-co-propylene) rubber (or c-EPR) and diisodecyl adipate (or DIDA). Arrows illustrate the procedure used for initializing a phase separation of the solution of 12wt% polymer.

system when it is cooled. Modulated structure due to the spinodal decomposition (SD) is observed. Then, the solvent-rich phase starts to appear as holes and the holes grow with time. After that, the phase separation manner is changed from a fluid-like mode to a gel-like mode. With the growth of solvent holes, the polymer-rich phase becomes network structure (1-4). Finally, when temperature reaches the crystallization point of the polymer, the phase structure is frozen. If the film thickness is suitably selected, and the resultant material is a two-dimensional continuous micro-reticulated network. Figure 2 shows a typical frozen phase pattern of the solution in a thin film of thickness of about 8μm, after being cooled from 150 to 23°C at a cooling rate of 10°C /min. The resultant structure is a polymer-rich skeletal network. The structure is quite stable at room temperature and the material is typically elastic.

The formation of the two-dimensional continuous network of the polymer-rich phase is generally transient in nature, though the transient state may last for *hours* in common experimental time-scale. Without the crystallization, the super-network structure will eventually collapse in late stages of the phase separation, as driven by the free energy minimization of the system. Hence, the phase separation structure obtained in Figure 2 represents only an early period (i.e., 10 minutes) of the phase separation. The resultant material is a micro-reticulated network that can have a very dense distribution of uniform cells. Varying the cooling rates can control the cell sizes (see Figure 3).

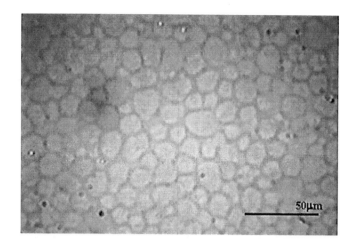

Figure 2. Phase pattern or structure of a 12wt% c-EPR/DIDA solution in a film of thickness of about 8μm, after being cooled from 150°C to 23 °C at a cooling rate of 10°C /min.

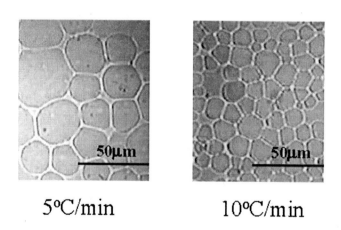

5°C/min   10°C/min

Figure 3. Control of cell size by change of cooling rate.

The evolution of network structures or patterns from the polymer solution at isothermal conditions is fascinating. To show this, the phase evolution on planner films was taken at 60°C isothermally, which is 45°C below the phase separation temperature 105°C, but 17°C above the melting temperature 43°C of the polymer in the solution. For films of thickness $L \geq 8\mu m$, just after a temperature quench, the sample first becomes cloudy and then microscopic solvent holes come into view (see Figures 4a and 4b). The number and the size of the holes increase with time, and some of them then start to coalesce. The polymer-rich phase accordingly becomes network-like structure after about three minutes (Figures 4c and 4d). The formation of the network usually takes less than 1 minute, but the network can persist for hours at 60°C. The network structure is very similar to that obtained at a constant cooling rate.

The kinetic of the cell growth, however, does not follow at constant growth rate (see Figure 5), when the solution is confined in the thin films. In the case of the nucleation-growth period, the phase domain is very small and the growth of the cell size (D) is given by a liner relation: $D \sim t$, where t is the holding time. After the cell size passes the film thickness, the growth speed of the cell size is considerably suppressed, and the growth of the cell size (D) is given by a power relation: $D \sim t^{1/3}$. Such a self-suppression of growing speed can support to give a network of uniform structure. Apparently, the kinetics of the cell growth is also influenced by the isothermal temperature. The growth rate at 90°C is systematically lower that that at 60°C.

Since the phase separation is driven by the free energy minimization of the system, the driving-force can therefore be expressed as a function of the solubility parameters (SP) of the two components in the solution, or particularly as a function of the difference, ΔSP. For the system investigated here, the SP value for the polymer is about 16.3 in terms of $(J/cm^3)^{1/2}$, and the SP value for the solvent DIDA is about 17.26. The difference between them is about 1.06. In order to show the effects of ΔSP on the phase separation, we have selectively studied several solvents, including tri-2-ethylhexyl trimellitate (TOTM, SP=18.3), di-2-ethylhexyl phthalate (DOP, SP=18.20), tri-2-ethylhexyl phosphate (TOP, SP=16.8), and paraffin oil (SP=16.4). The results are presented in Figure 6. In general, at constant temperature the larger the difference between the solvent and polymer SP values, the fast the oil-rich phase develops, and the bigger the cell size becomes at a given time.

The film thickness also strongly influences on the phase structure. Figure 6 presents the microscope observations on a wedge-like film of the polymer solution. The thickness of the film varies from 0 to $10\mu m$ over the wedge length of about 25 mm. After being quenched from 150°C to 60°C, hold isothermally for 10 minutes, and then being cooled down at a cooling rate of 10°C /min, the film is examined under a microscope and pictures are taken at various positions. When the thickness of the film is smaller than $L_c \cong 0.8\mu m$, phase separation

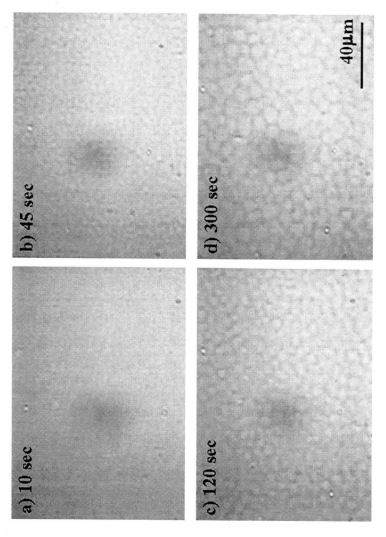

Figure 4. Phase evolution in films at 60°C for the c-EPR/DIDA 12 wt% solution in thin films after quenched from 150 to 60 °C.

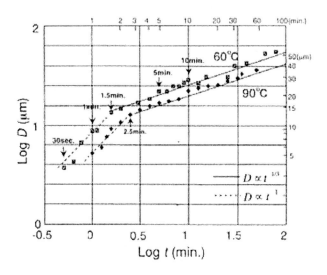

Figure 5. Time evolution of cell size as a function of time for the c-EPR/DIDA 12 wt% solution in thin films after quenched from 150 to 60°C and 90°C.

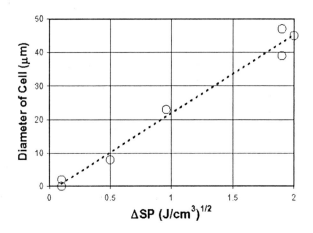

Figure 6. Effect of solvent quality on microstructure of a 12 wt% c-EPR/DIDA solution in a wedge-like film geometry films after quenched from 150 to 60°C, and held isothermally for 6 minutes

cannot be observed under the optical microscope. Above this critical thickness L_c, the film shows some phase separation, but the initial morphology appears to be very small polymer-rich droplets (see Figure 7a). As the thickness increases, the droplets increases in size and the morphology then gradually change to chains (see Figure 7b), and then the chains are gradually transformed to branches (see Figures 7c). The size of branches gradually grows as the thickness reaches to a critical value, i.e., $L_c^* \cong 4$ μm. Above the second critical thickness L_c^*, the branches are linked together (see Figure 7d). For films of thickness L>6 μm the phase structure returns back to the reticulated network that looks like the morphology in Figure 2. A schematic drawing of the phase morphology with increase of the film thickness is illustrated in Figure 8. Again, as L approaches $L_c \cong 0.8$ μm) from below, the film shows phase separation, but the initial morphology appears to be droplets. For L>L_c, the droplet size increases with increasing the film thickness. If further increasing the thickness L for L>L_c, we see a phase morphological transformation from droplets to chains, branches, and then to networks, while the concentration of the polymer stays almost constant for the change of thickness. There is correspondingly a critical thickness, L_c^* ($\cong 4$ μm > L_c), below which no bi-continuous phase separation takes place.

The existence of critical thickness in the phase separation process is surprising, especially under confinement of micrometer sizes. Our first speculation of the reason could be due to the special geometry of the wedge-like film. We therefore prepared a number of planar films in which one surface was in contact with a microscope glass slide but the other surface was exposed to air. Nevertheless, the same phase separation structures in wedge-like films are observed in the planar films. Again, for films of thickness L> L_c^*, the polymer-rich phase is a reticulated network structure after cooling. While, for films of thickness L < L_c^*, only droplets and branched chain structures are presented in the mixture. In addition, we also prepared planar films on hexamethyldisilazane-treated glass slides. The surface hydroxyl groups of the slides have been converted to trimethysilyl groups, which should provide better wetting for organic solutions. However, the same morphological transformation was observed. Accordingly, the de-mixing behavior for the binary system in confined slabs, as discussed above, cannot be interpreted in terms of the influences from a wedge-like geometry or from the glass surfaces, though the influence of surfaces on the phase separation process has become an important issue in material science (25-30).

Our second speculation of the reason for the anomalous phase transformation could be due to the transient nature of phase structures. The droplet morphology might be the early state of the network morphology, and the crystallization of the polymer might have frozen the phase evolutions into different states. To answer this question, we have investigated the phase

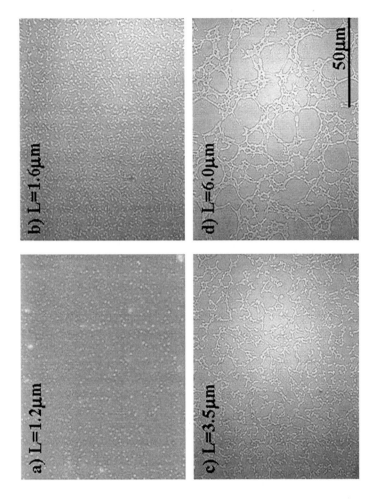

Figure 7. Phase structure of a 12 wt% c-EPR/DIDA solution in a wedge-like film geometry films after quenched from 150 to 60°C, and held isothermally for 10 minutes

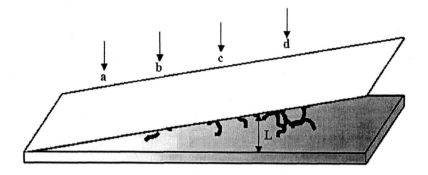

Figure 8. A schematic drawing of the morphological change observed in a wedge-like film geometry

evolution on planner films at 60°C isothermally. For films of thickness $L \gg L_c^*$ (i.e., $L > 8\mu m$), just after a temperature quench, the sample first becomes cloudy and then microscopic solvent holes come into view. The polymer-rich phase accordingly becomes network-like structure. The formation of the network usually takes less than 1 minute, but the network can persist for hours at 60°C. As time increases, the network eventually breaks into pieces, which then round up and collapse into big islands[4]. However, for films of thickness $L_c < L \ll L_c^*$ (i.e. $L \cong 1$~2 μm), phase separation starts with the formation of polymer-rich phase droplets after the temperature quench. The droplets once created are almost uniform in size and are moving freely by Brownian motion. Some of the droplets then stick together to form chains and branches, but frequently they break-up again. An important fact is that this droplet morphology is very stable and can last for a day at 60°C. After that, slowly some of the droplets coalesce together and form big islands. The droplet morphology in fact is more stable than the network morphology. It is therefore unlikely that the droplet morphology is the early state of the network morphology or vice versa. Our observation also suggests that phase separation in bulk and thick films starts with formation of solvent holes. While, in thin films the phase separation starts with formation of droplets of polymer molecules.

Finally, what is the role of the polymer concentration that plays in the phase morphology transformation observed in thin films? To answer this question, we have investigated a number of solutions of various polymer concentrations ranging from 4.5 to 12 wt%. Nevertheless, the same morphological transformation is observed in all those solutions as long as they are confined in the thin film geometry. The only difference is that, as polymer concentration increases, the solution viscosity increases and the time for phase separation increases. Our experiments also extend further on the phase separations of non-crystalline poly(ethylene-co-propylene) rubbers (or EPRs) in the diisodecyl adipate (or DIDA) solvent. Again, the phase separation for those non-crystalline systems shows the same morphological transformation in the thin film geometry. To the best of our knowledge, this is the first time that experiments provide such clear evidence that phase separation is very sensitive to micrometer confinements.

Summary

We developed a new technology, namely the frozen-in spinodal decomposition (FISD) (*1-5, 10-11*), to create two-dimensional polymeric skeleton-like reticulated networks of cell sizes of few micrometers. In this study, a small amount of a semi-crystalline polymer (~ 10 wt%) was first mixed with a low molecular weight solvent at an elevated temperature to obtain a homogeneous solution. Then, the homogeneous binary mixture was coated as a thin film, and was allowed to cool or quench into the unstable part of its miscibility gap. As a result, a spontaneous growth of the long wavelength concentration fluctuations (i.e., spinodal decomposition) took place in the system, and a bi-continuous phase morphology was formed. The phase pattern was then preserved or *frozen* by the crystallization of the polymer in the polymer-rich phase upon further cooling. The key principle in the technology is the fine-tuning of the solvent quality for that the crystallization of the polymer occurs just below the miscibility gap. The resultant material is a micro-reticulated network that can have a very dense distribution of uniform cells. Varying the cooling rates, holding time, solvent quality, and the compositions can control the cell sizes. After removing the low molecular weight solvent from the polymeric reticulate structure, there is also a two-dimensional continuous network skeleton that is consisting essentially of the pure polymer. This network is capable of effectively trapping a variety of other low molecular weight materials and has found a wide variety of applications, such as in liquid crystal displays, battery cell separators, membrane filters, and high damping additives. In addition to the technological importance, the *frozen-in* technology also provides the opportunity in fundamental studies of spinodal decomposition processes. Especially, the technology can freeze any moment of the evolution in the phase separation for detail investigations.

References

1. Wang, X. and Mashita, N. *Polymer*, **2004**, 45, 2711.
2. Fukahori, Y. and Mashita, N. *Polym. Adv. Technol.* **2000**, 11, 472.

3. Mashita, N. and Fukahori, Y. *Polymer Journal*, **2000**, 34(10), 719.
4. Fukahori, Y and Mashita, N. *Polym. Prepr. Jpn.*, **1995**, 44, 1738.
5. Mashita N. and Fukahori, Y. *Kobunshi Ronbunshu* **2000**, 57 (9), 596.
6. Tanaka, H and Nishi, T. *Japan, J. Appl. Phys.* **1988**, 27, L1787.
7. Tanaka, H. *Macromolecules* **1992**, 25, 6377.
8. Tanaka, H. and Miura T. *Phys. Rev. Lett.* **1993**, 71, 2244.
9. Tanaka, H. *J. Chem. Phys.* **1994**, 100, 5323.
10. USP5451454, USP5716997, EP0699710A2, USP5910530, USP6048930.
11. JP3310108, JP3336648, JP3298107, JP3376715, JP3378392.
12. Tanaka, H. *Phys. Rev. Lett.* **1996**, 76, 787.
13. Hashimoto, T.; Itakura, M.; and Hasegawa, H. *J. Chem. Phys.* **1986**, 85, 6118.
14. Bates, F. S. and Wilzius, P. *J. Chem. Phys.* **1989**, 91, 3258.
15. Binder, K. *Adv. Polym. Sci.* **1991**, 112, 181.
16. de Gennes, P. G. *J. Chem. Phys.* **1980**, 72, 4756.
17. Pincus, P. *J. Chem. Phys.* **1981**, 75,1996.
18. Onuki, A. *J. Chem. Phys.* **1986**, 85,1122.
19. Kawasaki, K. *Macromolecules* **1989**, 22, 3063.
20. Kotnis, M. A. and Muthukumar, M. *Macromolecules*, **1992**, 25, 1716.
21. Larson, R. G. *The structure and Rheology of Complex Fluids*, **1999**, Oxford,Oxford Univesity Press.
22. Haas, C. K. and Torkelson, J. M. *Phys. Rev. Lett.* **1995**, 75, 3134.
23. Tanaka, H. *Phys. Rev. E.* **1997**, 56, 787.
24. Tanaka, H. *J. Phys.: Condens Matter*, **2000**, 12, R207-264.
25. Fischer, H. P.; Maass, P.; Dieterich, W. *Eurphys Letts*, **1998**, 42(1), 49.
26. Ball, R. C. and Essery, R. L. H. *J. Phys.: Condens. Matter*, **1990**, 2, 13303.
27. Jones, R. A. L.; Norton, L. J.; Kramer E. J.; Bates, F. S.; and Wiltzius, P. *Phys. Rev. Lett.* **1991**, 66, 1326.
28. Bruder, F. and Brenn, R. *Phys. Rev. Lett.* **1992**, 69, 624.
29. Fischer, H. P.; Maass, P.; and Dieterich, W. *Phys. Rev. Lett.* **1997**, 79, 893.
30. Jinnai, H.; Kitagishi, H.; Hamano, K.; Nishikawa, Y.; and Takahashi, M. *Phys. Rev. E.* **2003**, 67, 21801.

Chapter 10

Fabrication of Conductive Au-Linker Molecule Multilayer Film on Flexible Polymer Substrates

Lakshmi Supriya[1] and Richard O. Claus[1–3]

[1]Macromolecular Science and Engineering Program, [2]Department of Electrical and Computer Engineering, and [3]Department of Materials Science and Engineering, Virginia Polytechnic Institute and State University, Blacksburg, VA 24061

Solution-based methods for the deposition of conductive Au film on flexible polymer substrates such as Kapton and polyethylene have been developed. The polymer is first treated in a plasma to activate the surface, subsequently silanes with terminal functional groups such as $-SH$, $-NH_2$, and $-CN$ that have an affinity for Au are deposited from solution. Multilayer Au films are fabricated by a layer-by-layer assembly process in which the silane coated sample is alternately immersed in a solution of colloidal Au and a bifunctional linker molecule. The films were characterized by atomic force microscopy (AFM), scanning electron microscopy (SEM), UV-vis spectroscopy and X-ray photoelectron spectroscopy (XPS). The electrical properties of the film were measured by a four-point probe and a setup to measure current and voltage. The resistance of the multilayer films was observed to be a function of the length of the linker molecule used to build the layers. The intitial coverage of Au particles on the surface was found to be affected by the particle size and concentration. The smallest sized particles with a high concentration gave the best coverage.

Introduction

Materials composed of two-dimensional (2-D), and three-dimensional (3-D) assemblies of nanoparticles and colloids are gaining increased importance in the field of nanoparticle-based devices (*1-5*). The properties of colloidal particles are usually in between those of the bulk material and the single atom, and these materials offer the advantage of easy tailorability of the surface with a change in the properties of the individual particles. Tuning the particle size, shape, composition, coverage on surface, and linking chemistries leads to different material properties (*6-8*). Au and Ag colloids have been the subject of extensive research in the past few years for the fabrication of films with different optical and electrical properties (*9-11*). The applications of these assemblies in substrates for surface plasmon resonance (*12*), and surface-enhanced Raman spectroscopy (*9*), have been extensively studied. Apart from this, they have also been studied as catalytic surfaces (*13*), and for biosensing applications (*14*).

Most of the previous work on colloidal assemblies has been done on rigid substrates such as glass and silicon. With the rapid emergence of plastic electronics, the need for controlled assemblies on flexible substrates has grown. In this paper, a solution-based approach for the formation of conductive Au electrodes on flexible polymer substrates and the effect of particle size on coverage have been described. The attachment of Au particles on polymer substrates requires the presence of groups that have an affinity for Au, like –SH, –NH$_2$, and –CN. This can be achieved by deposition of materials that have these groups at the terminals. The deposition of silanes with terminal functional groups is a convenient route for the attachment of Au (*15*). However, the deposition of silanes requires the presence of groups like hydroxyl on the surface. The silanes react with these hydroxyl groups in the presence of moisture to form covalent siloxane linkages. The surface functional groups are already present in surfaces such as glass or silicon, but for polymer surfaces they have to be created. This can be achieved by plasma treatment of the polymer surface (*16*). Plasma treatment creates free radicals, which when exposed to air get oxidized and can then attach to the silane. Immersion of the functionalized polymer film in colloidal Au leads to the assembly of Au particles on the surface.

Conductive films of Au can be fabricated by different approaches. Brown et al. (*17*) have developed a "seeding" process which involves the reduction of Au onto the Au particles already attached to the surface. Other methods include formation multilayer films of Au colloids and bifunctional linker molecules (*11, 18*), and formation of films from monolayer protected clusters of Au (*19*). In the formation of multilayer films, the electrical properties of the film were found to be a function of the length of the linker molecule, providing another factor to control the properties. In this paper the linker molecule approach has been used to fabricate conductive Au films on polymer substrates. The effect of particle size and concentration on the film properties has also been studied.

Experimental

Surface Modification of Polymer Films

The polymer used, Kapton (DuPont, 5 mil thickness), was treated in argon plasma in a March Plasmod plasma etcher for about 2 min, at 0.2 Torr and 50 W power. The treated polymers were immediately dipped in 1% (v/v) solution of 3-aminopropyl trimethoxy silane (APS), or 3-mercaptopropyl trimethoxy silane (MPS) in methanol for 10 min with stirring at room temperature. Then the substrates were rinsed thoroughly in methanol, dried in a stream of N_2 and heated at 110°C for 90 min to complete siloxane formation.

Synthesis of Colloidal Gold

The complete procedure for the synthesis has been described elsewhere (20). Briefly, to a rapidly boiling solution of sodium citrate (106 ml, 2.2 mM), 1 ml of 24.3 mM HAuCl$_4$ was slowly added with stirring. The reaction is complete in 2 min, when a wine red solution is obtained. The solution was boiled for further 15 min and was stored at 4°C. Different sizes of particles were made by changing the molar ratio of citrate to gold (21). To 24.3 mM HAuCl$_4$, 0.21 ml, 0.3 ml, and 0.5 ml citrate solution were added, which led to the formation of particles of different sizes. For making different concentrations of Au solutions, the amounts of materials used were as follows: 2 times–50 ml, 4.4 mM citrate and 1 ml, 24.3 mM HAuCl$_4$, 3 times–50 ml, 4.4 mM citrate and 1.42, ml 24.3 mM HAuCl$_4$, 4 times–100 ml, 8.8 mM citrate and 2 ml, 24.3 mM HAuCl$_4$, 6 times–100 ml, 13 mM citrate and 5.68 ml, 24.3 mM HAuCl$_4$.

Formation of Au Film

For the multilayer deposition, the silane coated polymer was immersed in colloidal Au solution for 1 h, and then rinsed in water, immersed in the bifunctional linker molecule solution for 15 min, rinsed in water and repeated to get the desired number of bilayers. The linker molecules used were, 2-mercaptoethanol (ME), 4 mm solution in water, 1, 6-hexanedithiol (HD), 4 mM solution in ethanol, and 1, 10-decanedithiol (DD), 4 mM solution in ethanol. For the studies on the effect of particle size on film properties, the substrates were immersed in colloidal Au solution for various times ranging from 1 to 24 h. The UV-vis studies on the films were performed on silane coated quartz slides.

Characterization

XPS spectra were acquired on a Perkin-Elmer 5400 X-ray photoelectron spectrometer. The spectra were corrected with reference to the C 1s peak at 285.0 eV. The gold solutions were characterized by UV-vis spectroscopy using a Hitachi U-2001 spectrophotometer and a Philips 420T transmission electron microscope (TEM). The Au films were examined by a Leo 1550 field emission SEM operating at an accelerating voltage of 5 kV and a Nanoscope IIIa AFM.

Results and Discussion

Fabrication of Gold Films

The fabrication of conductive gold films on flexible polymer substrates using the "seeding" approach has been described previously (20). Figure 1 illustrates the scheme for the fabrication of multilayer films using Au colloid and a linker molecule. The effective plasma treatment of the polymer and silane deposition was characterized by XPS. A Si 2p peak at 102.4 eV confirmed the siloxane formation, as well as the presence of the –SH or –NH$_2$ on the polymer surface after silane deposition. The first immersion in Au solution causes the adsorption of Au particles on the surface. Further immersions in the linker molecule solution and Au solution alternately, lead to additional Au adsorption. Figure 2(a) is the SEM image of the gold coverage after a single immersion in a colloidal Au solution and figure 2(b) shows the gold coverage after 10 bilayers of Au-ME. (A bilayer refers to the layers formed upon a single immersion in a Au solution and subsequent immersion in the linker solution). From the images it is very evident that there is increase in the adsorption of Au particles upon alternate immersion in the two solutions. After 10 bilayers all the films acquire the shiny appearance of bulk gold.

The growth of the film was followed by UV-vis spectroscopy. Spectra were measured on a quartz slide derivatized with silane and immersions in Au colloid and linker molecule solutions. Figure 3 shows spectra obtained for 2, 4, 6, 8, and 10 bilayers of Au-HD. A continual increase in absorbance is observed with the addition of the layers. There is also a slight red-shift in the peaks with the addition of layers. The absorption peak for 2 bilayers is at 530 nm, while it increases to 564 nm for 10 bilayers. This red-shift is attributed to the increased aggregation of the particles. These results agree well with the results obtained by Musick et al. (11). Similar results were observed for multilayers formed from ME and DD.

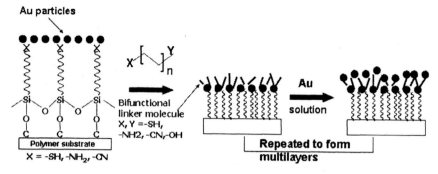

Figure 1. Schematic illustrating the layer-by-layer assembly process of Au multilayer films.

It has been observed that the electrical properties of the multilayer film depend greatly on the length of the linker molecule (11). The smaller the length of the linker, the better the conductivity. Similar behavior was also observed on the films on polymer substrates. A 10 bilayer film of Au-ME had a resistance of ca. 30 Ω, while there was a dramatic increase in the resistance for a similarly formed Au-HD film, which was found to be 1.5 MΩ. A 10 bilayer Au-DD film was not conductive at all and deposition of 15 bilayers did not make it conductive. This behavior has been discussed previously (11), however a conclusive reasoning for the trend has not been developed. Xu and Tao (22) have measured the resistance of the individual HD and DD molecules and found that to be ca. 10 MΩ and ca. 630 MΩ, respectively. This increased resistance of longer chain molecules coupled with the possibility that longer chain molecules separate the Au particles further, eliminating a conductive pathway, maybe the reason for the increased resistance. Some evidence for this hypothesis is provided by our current studies on the thermal effects on resistance, where the resistance decreases upon desorption of the linker molecule (23). This method of fabricating conductive films on flexible substrates with the ability to control the resistance, provides an inexpensive way for fabricating electrodes for plastic electronic devices.

Effect of particle size

For the successful fabrication of electronic devices, a good control over the electrode properties needs to be achieved. During the fabrication of the electrodes as described above, and also by the "seeding" method, it was

142

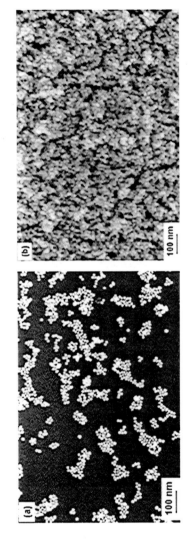

Figure 2. SEM image of Au on Kapton a) after single immersion in Au solution, b) after 10 bilayers of Au-ME.
((a) Reproduced with permission from Langmuir, **2004**, 20, 8870-8876. Copyright 2004 American Chemical Society.)

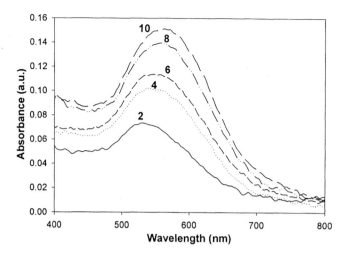

Figure 3. UV-vis spectra depicting the growth of the Au-HD film. The numbers indicate the number of layers.

observed that the amount of Au adsorbed after the first immersion and the surface coverage played an important role in the final electrical properties of the film. A low initial surface coverage led to higher final resistance. This effect was more prominent in the electrodes fabricated by the "seeding" approach. This prompted the investigation into the factors affecting the initial surface coverage. Different sized Au colloidal particles were synthesized using a previously described method by Frens (21). By changing the molar ratio of citrate to gold salt, different sizes were obtained. This ratio was changed by keeping the amount of gold same and adding different volumes of the citrate. The particle sizes were measured by TEM and the solutions were characterized by UV-vis spectroscopy. As the size of the particles increased, a red-shift and broadening of the spectra was observed. The broadening of the spectra is attributed to the increasing polydispersity of the colloids as the size increases. It was observed from the TEM images also, that as the size of the particles increases, they become more elliptical in shape and also the sizes are not all the same. Table 1 shows different properties of the colloidal particles synthesized.

The silane coated polymers were immersed in the different Au solutions for 24 h. After that they were rinsed thoroughly in water and dried in a stream of nitrogen. The adsorption of the gold was characterized by AFM. Figure 4 shows the AFM image of the samples coated with different sized Au. From the images it is observed that the coverage is least for the biggest size particles. There are a lot of uncovered regions visible. As the size of the sample decreases the surface coverage increases and the surface roughness decreases. Grabar et al. (24) have reported that the Au adsorption of silane coated glass is diffusion controlled in

144

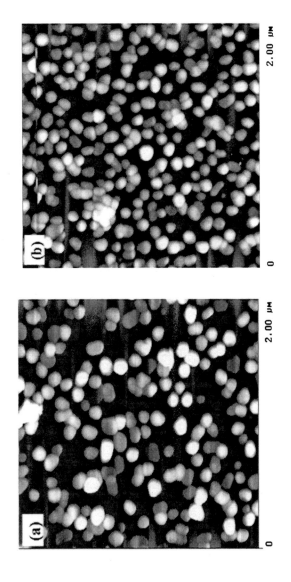

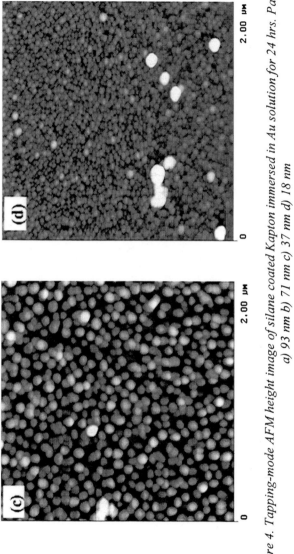

Figure 4. Tapping-mode AFM height image of silane coated Kapton immersed in Au solution for 24 hrs. Particle sizes a) 93 nm b) 71 nm c) 37 nm d) 18 nm

Table 1. Properties of the Au solutions.

Amount of citrate (ml)	Size (nm)	Ellipticity	λ_{max} (nm)	FWHM (nm)
0.21	93	1.33	556	154
0.3	71	1.32	544	132
0.5	37	1.16	535	82
1.0	18	1.10	523	82

Note: Ellipticity is the ratio between the two diameters of an ellipse

the first hour of immersion and beyond that the coverage is limited by the repulsive interparticle interaction (Au particles synthesized by this method are negatively charged). Another study has observed similar behavior on silane coated glass and has stated that the larger particles are less effectively immobilized on the silane surface initially leading to low coverage and more aggregation (25).

Effect of particle concentration

Another factor that affects the surface coverage is the concentration of particles in solution. Solutions of different concentrations were made by keeping the ratio of gold to citrate same, but increasing the amounts in a given volume of solution. The size of the Au particles used was 18 nm. The concentrations were increased 2, 3, 4, and 6 times the standard solution. The solutions were characterized by UV-vis spectroscopy. The peak absorbance for all solutions was at 522 nm indicating that all solutions have particles of the same size, although the absorbance increased with the increasing concentration. This evident by visual inspection of the solutions also; as the concentration increases, the color changed from a light wine-red to a deep red. The silane coated polymers were immersed in the different solutions for 24 hrs, then rinsed in water and dried. AFM images of the sample show good particle coverage for all samples with a slightly increased coverage for the most concentrated solution. The roughness for all the samples is less than 4 nm, indicating very smooth surfaces. The roughness was found to decrease to about 1.5 nm for the 4 times concentrated solution and then increased to about 3.9 nm for the most concentrated solution. This is possibly due to the increased amount of Au being adsorbed which starts forming small clusters.

Conclusions

A solution-based approach for the formation of conductive multilayer Au film on flexible polymer substrates has been developed. This requires the functionalization of the polymer surface by silanes with terminal groups like – SH, and –NH2 and alternate immersion in colloidal Au solution and linker molecule solution. Three bifunctional molecules with different chain lengths were used as the linker molecule. It was observed that the resistance of the film increased dramatically with increasing chain length. For the 10-C chain linker, even after depositing 15 bilayers the film was insulating. The Au coverage on the polymer was found to be affected by the particle size and particle concentration. The coverage was best for the smallest particle with the highest concentration. This initial coverage is critical as it influences the final film properties.

Acknowledgements

Financial aid from the U.S. Army Research Laboratory and U.S. Army Research Office under contract/grant number DAAD19-02-1-0275 Macromolecular Architecture for Performance (MAP) MURI is acknowledged. We also thank Stephen McCartney, Materials Research Institute for help with the SEM images, Frank Cromer, Department of Chemistry Virginia Tech for the XPS data and Kim Donaldson, Department of Materials Science and Engineering, Virginia Tech for help with the AFM images.

References

1. Walt, D. R. *Nature Materials* **2002**, *1*, 17-18.
2. McConnell, W.; Brousseau, L. C., III; House, A. B. ; Lowe, L. B.; Tenent, R. C.; Feldheim, D. L. *Metal Nanoparticles* eds. Feldheim, D. L.; Foss, C. A., Jr. Marcel Dekker, Inc., New York, NY, **2002**, 319-333.
3. Shenhar, R.; Rotello, V. M. *Acct. Chem. Research* **2003**, *36*, 549-561.
4. Willner, I.; Willner, B. *Pure and Appl. Chem.* **2002**, *74*, 1773-1783.
5. Rao, C. N. R.; Cheetham, A. K. *J. Mater. Chem.* **2001**, *11*, 2887-2894.
6. Xu, X.-H, N.; Huang, S.; Brownlow, W.; Salatia, K.; Jeffers, R. B. *J. Phys. Chem. B* **2004**, *108*, 15543-15551.
7. Halperin, W. P. *Rev. Mod. Phys.* **1986**, *58*, 533-607.
8. Stephan, L.; Mostafa, A. E.-S. *Int. Rev. Phys. Chem.* **2000**, *19*, 409- 453.
9. Grabar, K. C.; Freeman, R. G.; Hommer, M. B.; Natan, M. J. *Anal. Chem.* **1995**, *67*, 735-743.

148

10. Brust, M.; Kiely, C. J. *Colloids and Colloidal Assemblies* ed. Caruso, F., Wiley-VCH Verlag GmbH & Co. KGaA, Weinheim, Germany **2004**, 96-119.
11. Musick, M. D.; Keating, C. D.; Lyon, L. A.; Botsko, S. L.; Pena, D. J.; Holliway, W. D.; McEvoy, T. M.; Richardson, J. N.; Natan, M. J. *Chem. Mater.* **2000**, *12*, 2869-2881.
12. Jin, Y.; Kang, X.; Song, Y.; Zhang, B.; Cheng, G.; Dong, S. *Anal. Chem.* **2001**, *73*, 2843-2849.
13. Jin, Y.; Shen, Y.; Dong, S. *J. Phys. Chem. B* **2004**, *108*, 8142-8147.
14. Natan, M. J.; Lyon, L. A. *Metal Nanoparticles* eds. Feldheim, D. L.; Foss, C. A., Jr. Marcel Dekker, Inc., New York, NY, **2002**, 183-205.
15. Goss, C. A.; Charych, D. H.; Majda, M. *Anal. Chem.* **1991**, *63*, 85-88.
16. Rose, P. W.; Liston, E. M. *Plastics Engg.* **1985**, *41(10)*, 41-45.
17. Brown, K. R.; Lyon, L. A.; Fox, A. P.; Reiss, B. D.; Natan, M. J. *Chem. Mater.* **2000**, *12*, 314-323.
18. Musick, M. D.; Keating, C. D.; Keefe, M. H.; Natan, M. J. *Chem. Mater.* **1997**, *9*, 1499-1501.
19. W. P. Wuelfling, F. P. Zamborini, A. C. Templeton, X. Wen, H. Yoon, R. W. Murray, *Chem. Mater.* **2001**, *13*, 87.
20. Supriya, L.; Claus, R. O. *Langmuir* **2004**, *20*, 8870-8876.
21. Frens, G. *Nature Phys. Sci.* **1973**, *241*, 20-22.
22. Xu, B.; Tao, N. J. *Science* **2003**, *310*, 1221-1223.
23. Supriya, L.; Claus, R. O. *J. Phys. Chem. B* **2005**, *109*, 3715-3718.
24. Grabar, K. C.; Smith, P. C.; Musick, M. D.; Davis, J. A.; Walter, D. G.; Jackson, M. A.; Guthrie, A. P.; Natan, M. J. *J. Am. Chem. Soc.* **1996**, *118*, 1148-1153.
25. Park, S.-H.; Im, J.-H.; Im, J.-W.; Chun, B. –H.; Kim, J. –H. *Microchem J.* **1999**, *63*, 71-91.

Chapter 11

Block Copolymer Nanocomposite Films Containing Silver Nanoparticles

K. Swaminathan Iyer[1], Jeff Moreland[1], Igor Luzinov[1,*],
Sergiy Malynych[2], and George Chumanov[2,*]

[1]School of Materials Science and Engineering and [2]Department
of Chemistry, Clemson University, Clemson, SC 29634

Thin nanocomposite films consisting of silver nanoparticles
(100 nm) evenly dispersed in poly(styrene-b-2-vinyl pyridine)
(PS-PVP) block copolymer were formed via melt
incorporation. Block copolymer films of 30-250 nm thick
were deposited on silicon and glass substrates by dip-coating
from solution. First, the nanoparticles were attached to the
PVP domains exposed to the surface of the block copolymer
film. Next, the substrates covered with nanoparticles were
annealed at 130°C in vacuum oven for different periods of
time. The thermal treatment caused incorporation of the
nanoparticles into the polymeric film. Level of the
nanoparticles incorporation strongly depended on the thickness
of the block copolymer film. The formation of the
nanocomposite films was monitored by scanning probe
microscopy, contact angle measurements, and different
spectroscopic techniques.

Introduction

Growing demand for the fabrication of electronic components, optical detectors and biochemical sensors with nano-dimensions has initiated theoretical and experimental investigations in the field of nanocomposite films. Nanocomposites of inorganic materials in organic matrices are of particular interest.[1] They combine the properties of polymers like elasticity and dielectric properties with the high specific surface of nanoparticles.[1] Metal particles, which constitute the inorganic inclusions in the organic matrices of the nanocomposite, comprise a fundamentally interesting class of matter because of an apparent dichotomy between their sizes and many of their physical and chemical properties.[2] Silver particles reduced to nanometer dimensions are especially interesting objects, since they exhibit unique optical properties in visible spectral range due to the excitation of the collective oscillations of conducting electrons known as plasmon resonance or surface plasmons.[3,4]

One of the major concerns during the fabrication of nanocomposite films is immobilization of these nanoparticles on dielectric surfaces. The immobilization is often accomplished by surface modification with functional groups that are attractive to the particles. Functional groups such as thiols are often used to immobilize metal nanoparticles on various oxide surfaces.[5] The thiol group forms strong bonds with the nanoparticles resulting in self-assembled structures with unique optical and electrical properties.[6] Polymers such as poly(vinylpyridine) PVP have been also used as effective adhesives to immobilize silver nanoparticles.[7] Each PVP macromolecule provided many binding sites for simultaneous interaction with the metal nanoparticle and the substrate. The cooperative interaction resulted in strong binding of the nanoparticles. Nanoparticles embedded in thin polymer films of poly(vinylalcohol) and poly(vinylpyrrolidone) have been produced in-situ by thermal annealing of polymer films containing metal salts.[1] The above-discussed routes are illustrated in Figure 1. Irrespective of the technique used for the fabrication of nanocomposites, the prime requirement for it to be functionally sound is that the nanoparticles should be stabilized on the surface and protected from mechanical and chemical effects.[8]

The strategy developed in the current study for the fabrication of the nanocomposite films has three basic steps (Figure 2): (a) deposition of polymer film on a substrate, (b) attachment of nanoparticles to the film surface and (c) incorporation (encapsulation) of the nanoparticles in the film by thermal treatment. To realize the proposed strategy, the polymer has to be carefully selected to spread over the nanoparticle surface to induce encapsulation at elevated temperature. The ability of the polymer to spread over the particle surface is given by its positive spreading coefficient $(S_{p/n})$[9]:

$$S_{p/n} = \gamma_n - \gamma_p - \gamma_{pn} > 0 \qquad (1)$$

where γ_p is the surface energy of the polymer, γ_n is the surface energy of the nanoparticle and γ_{pn} is the interfacial tension at the polymer/nanoparticle interface. Equation (1) implies that for the encapsulation of the nanoparticle to be successful, the surface energy of the polymer (γ_p) must be sufficiently lower than the surface energy of the nanoparticle (γ_n). This condition is represented by equation (2):

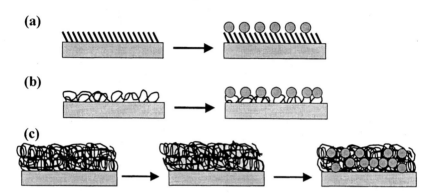

Figure 1. Routes for immobilization of nanoparticles. Attachment of nanoparticles to functionalized (a): self assembled monolayer; (b): adsorbed polymer layer and (c): Synthesis of nanopartciles inside polymer film through precursors.

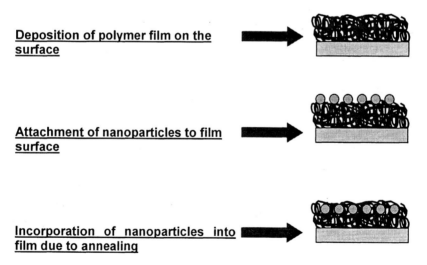

Deposition of polymer film on the surface

Attachment of nanoparticles to film surface

Incorporation of nanoparticles into film due to annealing

Figure 2. Strategy used for preparation of nanocoposite film.

$$\gamma_n > \gamma_p \tag{2}$$

On the other hand, γ_{pn} has to be adequately low for the spreading coefficient to be positive. However, if surface of an inorganic nanoparticle is polar (typical situation) the polymer should be also polar (high γ_p) for the later condition. Thus, it is quite difficult to pin-point a polymer that will satisfy those somewhat opposite requirements for γ_p. To this end, a diblock copolymer with a short polar block and a long non-polar block might be employed to resolve the interfacial conflict.

In the current study incorporation of silver nanoparticles into poly(styrene-b-2vinylpyridine) (PS-b-PVP) block copolymer thin film deposited on silicon and glass substrates was investigated. PVP was already reported[7] to be an effective surface modifier for immobilization of nanoparticles. The strong affinity of the pyridyl groups to metals through the metal-ligand interactions of the nitrogen atoms aids the PVP block to immobilize silver particles. PVP can also interact electrostatically in quaternized or protonated forms with charged surfaces.[10,11] In addition, PVP molecules are capable of simultaneously forming multiple hydrogen bonds with the silanol groups on the oxidized silicon/glass surface.[8] The long PS blocks with lower surface energy than PVP[12] would prefer location at the air/nanoparticle interface. Thus, the short polar block will segregate to the nanoparticle boundary while the long non-polar block will encapsulate the nanoparticle into the polymer film.

Experimental

ACS grade toluene and reagent alcohol (ethanol) were obtained from Acros Organics and were used as received. Highly polished single-crystal silicon wafers of {100} orientation (Semiconductor Processing Co) and glass were used as substrates. Silicon substrates were used for thickness measurements, while glass substrates were used for optical investigations. The substrates were first cleaned in an ultrasonic bath for 30 min, placed in a hot piranha solution (3:1 concentrated sulfuric acid/ 30% hydrogen peroxide) for 1h, and then rinsed several times with high purity water (18MΩ cm, Nanopure). After being rinsed the substrates were dried under a stream of dry nitrogen under clean room 100 conditions.

PS-b-PVP (PS block, M_n = 110000 and PVP block, M_n = 55000, PDI=1.1) was dip coated (Mayer Fientechnik D-3400) from toluene solution of different concentrations to vary the thickness of the block copolymer deposited. (The copolymer was synthesized at Centre Education and Research on Macromolecules, University of Liege, Belgium. Professor R. Jerome is director of the Centre). The lower and the upper limits of the dip coater were set to enable complete dipping of the substrate. The thickness of the deposited PS-b-PVP layers measured by ellipsometry were 31 ± 1 nm, 57 ± 1.5 nm, 112 ± 1 nm, 215 ± 3.0 nm and 252 ± 3.0 nm, respectively

The modified substrates were treated with ethanol. (Ethanol is a solvent for PVP, while it is a non-solvent for PS.) Treatment with ethanol ensured that the short PVP blocks were brought to the surface by preferential molecular rearrangement aiding in nanoparticle immobilization. Thickness of the ethanol treated layers was measured by ellipsometry. The thicknesses of the layers remained unchanged after the ethanol treatment.

The modified substrates were exposed to a suspension of silver nanoparticles (110 nm –130 nm in diameter) in deionized water (18MΩ cm, Nanopure) at low ionic strength overnight. Low ionic strength is required to maintain substantial long-range electrostatic repulsion between particles and consequently minimize clustering of the nanoparticles.[3] The exposure time was maintained constant for all the substrates to regulate the density of the particles on the surface. The resulting semiregular 2D array of nanoparticles deposited on the modified substrates was annealed in a vacuum oven overnight at 130 ^{0}C to enable encapsulation of the nanoparticles.

Ellipsometry was performed with COMPEL discrete polarization modulation automatic ellipsometer (InOmTech, Inc.) at an incidence angle of 70°. A three layer model (silicon substrate + silicon oxide layer + PS-b-PVP layer) was used to simulate experimental data. The refractive index of 1.55 was used to calculate the thickness of the PS-b-PVP layers. Original silicon wafers from the same batch were tested independently and used as reference samples for the analysis of modified substrates.

Scanning probe microscopy (SPM) studies were performed on a Dimension 3100 (Digital Instruments, Inc.) microscope. We used tapping mode to study the morphology of the films in ambient air. Silicon tips with a spring constant of 50 N/m were used to scan surfaces. Imaging was done at scanning rates in the range of 1-2 Hz. UV-vis absorption spectra were recorded on spectrophotometer UV-2501PC (Shimadzu).

Results and Discussion

Morphology of PS-b-PVP Films

SPM analysis of the dip-coated PS-b-PVP films revealed that the covering was uniform on both micro and nano scales (Figure 3). The block copolymer films possessing different thickness demonstrated virtually the same surface

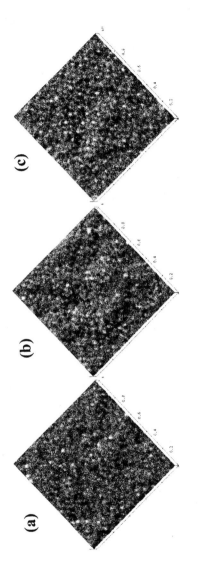

Figure 3. 1 x 1 μm SPM topography images of dip coated PS-b-PVP layers. Thickness (a): 31 nm, (b): 112 nm and (c): 252 nm respectively. Vertical scale: 11.25 nm.

morphology, where phase separation of the minor (PVP) component in spherical domains was clearly observed. However, a desired nanoparticle adsorption was not detected when the covered with the block copolymer film substrates were exposed to a suspension of silver nanoparticles. This implied that the low surface energy of the PS blocks resulted in preferential arrangement of the blocks at the air/copolymer interface due to condition of interfacial energy minimization. Irrespective of the thickness of the dip coated layer, the surface of the film primarily consisted of PS chains, which are inactive to nanoparticle adsorption.

The modified substrates were then treated with ethanol. (Ethanol is a solvent for PVP, while it is a non-solvent for PS.) SPM analysis of the ethanol treated surface (Figure 4) showed that the treatment with a selective solvent resulted in a "dimple" morphology. The morphology change resulted in the increase in the surface roughness of the ethanol treated film (Figure 5). The average SPM roughness of the dip-coated surface was 0.34 ± 0.02 nm, while that of the phase-segregated surface was 0.44 ± 0.06 nm. The dimples corresponded to PVP domains (compare Figures 3 and 4). The dimple formation revealed that the ethanol treatment caused the opening of the PVP domains.

Immobilization of Silver Nanoparticles

The PVP "nanopockets" (formed from the microphase-segregated structure after ethanol treatment) have ensured the nanoparticle adsorption. When the substrates were exposed to a suspension of silver nanoparticles, significant amount of the particles was accumulated on the topmost surface of the PS-b-PVP films. SPM analysis (Figure 6) confirmed the immobilization of the nanoparticles due to the affinity of pyridyl groups to silver through the metal-ligand interactions of the nitrogen atoms. The electrostatic repulsion between the particulates resulted in the formation of semi-regular two-dimensional nanoparticle arrays. (Multiple areas on the film surface were imaged and the structures shown in Figure 6 are representative of the entire film morphology.) The occasional rod shaped, [13] "dimers" or "trimers" constitute no more than few percent of the surface and in many areas none of these were observed. The average number of particles on a 3 μm x 3 μm area was 145 ± 10 for all the substrates irrespective of the PS-b-PVP film thickness. The fact that substrates with different amounts of block copolymer adsorbed the same amount of nanoparticles can be explained in terms of conformations adopted by the polymer molecules. The total number of pyridyl groups available for the attachment of nanoparticles at the surface reaches saturation provided that the entire substrate is covered with the polymer uniformly. An increase in the thickness of the block copolymer results in an increase in the number of pyridyl groups hidden in the interior of the film, without changing the amount of

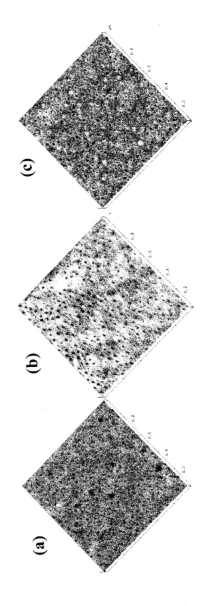

Figure 4. 1 x 1 μm SPM topography images of ethanol treated PS-b-PVP layers. Thickness (a): 31 nm, (b): 112 nm and (c): 252 nm respectively. Vertical scale: 11.25 nm.

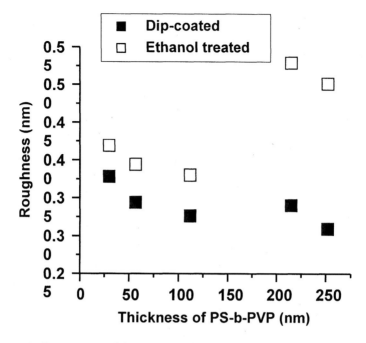

Figure 5. Comparison of the SPM roughness for the dip coated and ethanol treated surface of different thickness.

exposed groups available for interaction.[7,14] The particle elevation calculated using SPM images (Figure 6) was in the range 100 nm to 130 nm and this corresponded to the particle diameter for all substrates.

Two distinctive peaks characterize the excitation spectrum of colloidal suspension of 100 nm silver particles in water,[3] namely intense broad band at 545 nm due to dipolar component of the plasmon and a somewhat weaker band at 430 nm corresponding to the quadrapolar component. The excitation spectra of the particles assembled in a two-dimensional array on the substrates modified with PS-b-PVP are shown in Figure 7. The spectrum in water (Figure 7 (a)) is characterized by a single sharp peak at 459 nm, when compared to the two peaks of the colloidal suspension. This change is due to the change in the average interparticle distance in the array. When the average interparticle distance in the arrays become comparable to the particle dimensions, the

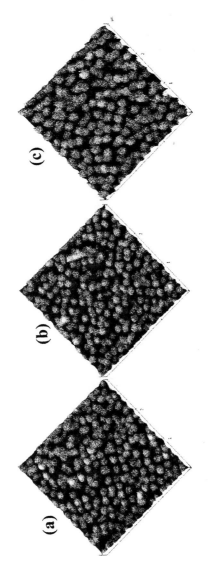

Figure 6. 3 x 3 μm SPM topography images of silver particles adsorbed on PS-b-PVP layers (dry state). Thickness (a): 31 nm, (b): 112 nm and (c): 252 nm respectively. Vertical scale: 225 nm.

excitation spectrum undergoes a dramatic change with the appearance of a single sharp band. The local fields from individual particles overlap when the particles are closely spaced. This near-field interaction results in the system being coupled. This single peak corresponds to a new optical mode called the cooperative plasmon mode (CPM).[3]

When the substrates were dried, silver particles underwent surface aggregation and the array lost its uniformity. The formation of random aggregates was confirmed by the spectra in dry state (Figure 7 (b)). Surface aggregation of the silver particles is characterized by broadening of the band and shift to the red spectral range due to strong coupling between particles.[15,16]

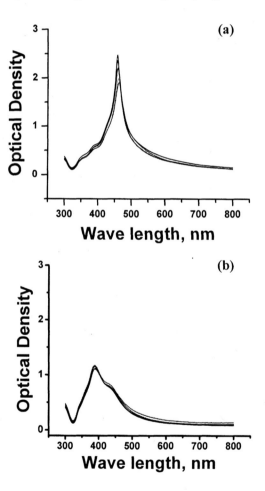

Figure 7. Excitation spectra of 100 nm Ag particles assembled in a closely spaced two-dimensional semi-regular array. (a): In water (b): After drying.

The peak was shifted to 390 nm after drying. An additional weak band also appeared at 445 nm, which corresponds to multipolar coupling modes depending on the degree of surface aggregation. The surface aggregation of the nanopartciles can be attributed to the capillary drag force of water.[17] Drying reduces the electrostatic repulsion between the particles and the surface tension of the evaporating water layer forces the nanoparticles to clump together. It was reported[7] that drying of substrates with solvent with smaller surface tension (e.g. alcohol) might reduce but not totally eliminate surface aggregation of silver nanoparticles.

Thickness of the block copolymer layer had no influence on the both spectra in water (Figure 7 (a)) and in the dry state (Figure 7 (b)). This was due to the fact that substrates with different amounts of block copolymer adsorbed same amount of nanoparticles that were arranged in a similar manner on the surface. The spectroscopic measurements provided supplementary to SPM imaging evidence that all polymer films had virtually the same structure of the particle arrays before annealing.

Encapsulation of silver nanoparticles

Annealing of the copolymer films covered with adsorbed silver nanoparticles (in a vacuum oven overnight at 130°C) dramatically changed the morphology of the films. SPM analysis of the annealed substrates reveled that the block copolymer film successfully incorporated the silver particles (Figure 8). Encapsulation of the silver nanopartciles into the block copolymer film is a direct effect of the block copolymer reorientation at elevated temperatures. (The domain orientation of a given block copolymer exhibits dependence on the energetic/chemical nature of the surface/interface that border the film.[18] This orientation is a result of surface and interfacial energy minimization in order to attain near equilibrium morphology.) For the nanocomposite system under consideration four kinds of interface, affecting the equilibrium film structure, are present: air/copolymer, air/silver, silver/copolymer, and substrate/copolymer.

Two synergistic processes at elevated temperatures control the encapsulation of the silver nanoparticles. One is related to the spreading of PS block due to its positive spreading coefficient ($S_{n/p}$). The second process corresponds to the surface diffusion of the PVP block around the silver nanoparticles at the silver/copolymer and at the substrate/copolymer interface due to the interfacial energy constraints. PVP block preferentially interacts at the silver/copolymer interface due to the strong affinity of the pyridyl groups to metals through the metal-ligand interactions of the nitrogen atoms and at the substrate/copolymer interface due to its capability of forming hydrogen bonds with the silanol groups on the oxidized silicon surface. The preferential

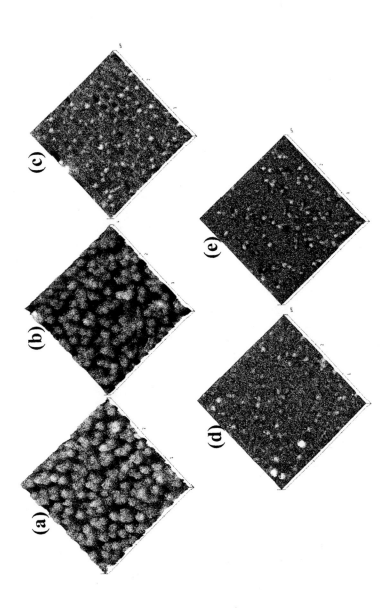

Figure 8. 3 x 3 μm SPM topography images of silver particles encapsulated in PS-b-PVP layers after annealing at 130°C. Thickness (a): 31 nm (b): 57 nm (c): 112 nm (d): 215 nm and (e): 252 nm respectively. Vertical scale: 225 nm.

interaction of PVP with silver and substrate, together with the drive for interfacial energy minimization will lead to a lamellar brush structure (oriented parallel to substrate) at the substrate/copolymer interface and a core shell brush structure (oriented concurrent to silver) at the silver/copolymer interface (Figure 9). The brushes may be characterized with two parts (typical to block copolymer brushes): the anchoring block (PVP) and the buoy block (PS). It was reported that a surface energy difference of as low as 0.1 dyn/cm between the blocks was sufficient to induce such reorientation.[19]

The SPM micro-roughness of the films was 42 nm, 27 nm, 7 nm, 5 nm and 6 nm for 31 nm, 57 nm, 112 nm, 215 nm and 252 nm thick films, respectively. It may be noted that the SPM roughness did not change significantly for films with thicknesses greater than 100 nm. Silver nanoparticle immersion in the block copolymer was estimated using SPM (Figure 10 (a)). For film thickness greater than 100 nm, immersion into the block copolymer matrix was fixed, confirming the claim for no significant value of the roughness. Static contact angle measurements were made on the annealed substrates with water (pH 7.0), and a static time of 30 seconds before angle measurement (Figure 10 (b)). The average contact angle for substrate before annealing was $68° \pm 2°$ and after annealing was $100° \pm 5°$. (Contact angles for pure PVP and PS films are 50-60° and 90°, respectively). The increase in contact angle after annealing was due to the PS blocks of the copolymer spreading over the silver particles and due to the increase in surface roughness as a result of nanoparticle adsorption.

The inability of the silver nanoparticles to penetrate beyond a certain extent into PS-b-PVP matrix can be explained in terms of thermodynamic limitations arising due to the core shell brush structure around the particle.[20] The core shell structure can be imagined to be a "buoy-anchor" assembly of the diblock copolymer around the silver nanoparticles with PVP anchor block and PS buoy block. The polymer chains around the nanoparticles are strongly anchored to the spherical silver substrate. Steric interactions between the "buoy-anchor" assembly and the neighboring polymer chains may induce short-range order in the vicinity of the silver particles. In order for the nanoparticle to fully immerse into the polymer matrix, the local structural order has to be disrupted. The requirement for the particle to immerse (disrupt local order) is to overcome a free energy barrier associated with the local structural order. This free energy barrier equilibrates the particle immersion in the block copolymer matrix.

Excitation spectrum of the annealed samples is shown in Figure 11. The spectral characteristic of the nanocomposite films strongly depended on the film thickness as the peak intensity decreased with the decrease in film thickness. Annealing resulted in a blue spectral shift (peak shifted from 390 nm to 483 nm) for thicker films while for film thickness less than the nanoparticle dimension (100 nm) annealing resulted in broadening of the peak. The blue spectral shift can be associated with thermal induced self-reparation in case of thicker films to

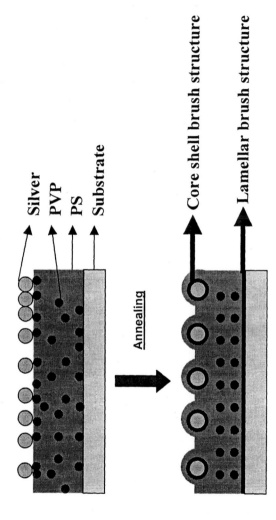

Figure 9. Schematic representation of the core shell structure and lamellar structure at the silver/copolymer interface and substrate/copolymer interface.

restore lost order during drying. In this case of the thicker films (with the thickness larger than the particle diameter) the particles are extensively incorporated into the block copolymer films. The PS brush surrounding a nanoparticle tends to get entangled with "unbound" PS chains of the block copolymer and, thus, have a tendency to repel the brush covering the neighboring particle. As the result of the repulsion the particles may maximize distance between them partially restoring the lost order. For the films with the thickness that is lower than the particle diameter (less than 100 nm thick)

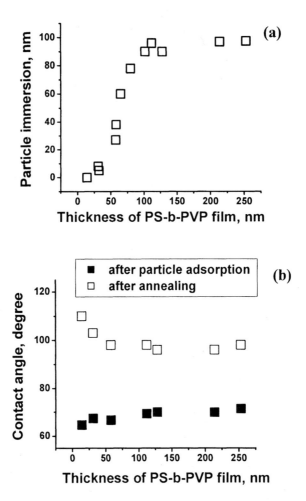

Figure 10. (a): Silver nanoparticle immersion in the block copolymer matrix for different thickness of PS-b-PVP. (b): Contact angle before and after annealing for different film thickness.

spectral broadening can be associated with additional thermally induced coalescing of nanoparticles. In this case a particle can not be immersed deeply into the film from geometrical considerations. Then from condition of minimization of interfacial tension the particles trying to decrease their interface with air and have tendency to minimize distance between them. This results in enhanced lateral mobility of the nanoparticles on the block copolymer film surface leading the particles to coalesce.

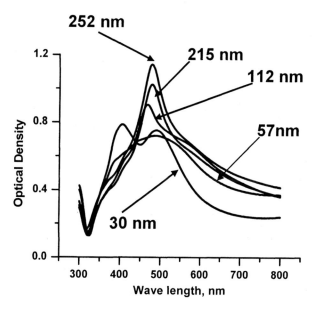

Figure 11: Excitation spectrum of silver encapsulated PS-b-PVP films after annealing at 130°C.

In summary, the current research provides a novel approach to fabricate organic/inorganic nanocomposites for possible biosensor, optical and electronic applications. PS-b-PVP block copolymer film is shown to be highly efficient in nanoparticle adsorption. Annealing of the silver/block copolymer system resulted in encapsulation of nanoparticles into PS-b-PVP matrix due to the difference in the surface energies. It is believed that the "anchor-buoy" assembly of the physisorped PS-b-PVP brushes stabilizes the nanoparticles in the copolymer matrix. The optical properties of the nanocomposite films depended on the initial thickness of the block copolymer film. The number of silver nanoparticles adsorbed by all substrates remained approximately the same irrespective of the film thickness, but the optical properties were regulated by the interparticle distance, which was controlled by the initial polymer thickness.

Acknowledgements This work is supported by Department of Commerce through National Textile Center, C04-CL06 Grant and by the ERC Program of National Science Foundation under Award Number EEC-9731680. The authors would like to thank Professor Dr. R. Jerome (University of Liege) for the providing with the block copolymer sample.

References

1. Fritzsche, W.; Porwol, H.; Wiegand, A.; Bornmann, S.; and Köhler, J. M. *Nanostructured Materials*, **1998**, 10, 89.
2. McConnell, W. P.; Novak, J. P.; Brousseau III, L. C.; Fuierer, R. R.; Tenent, R. C.; and Feldheim, D. L. *J. Phys. Chem. B* **2000**, 104, 8925.
3. Malynych, S.; and Chumanov, G. *J. Am. Chem. Soc.* **2003**, 125, 2896.
4. Kreibig, U.; Vollmer, M. *Optical Properties of Metal Clusters*; Springer Verlag: Berlin, Heidelberg, **1995**.
5. Goss, C. A.; Charych, D. H.; Majda, M. *Anal. Chem.* **1991**, 63, 85.
6. Chumanov, G.; Sokolov, K.; Cotton, T.M. *J. Phys. Chem.* **1996**, 100, 5166.
7. Malynych, S.; Luzinov, I.; and Chumanov, G. *J. Phys. Chem. B* **2002**, 106, 1280.
8. Malynych, S.; Robuck, H.; and Chumanov, G. *Nano Lett.* **2001**, 1, 647.
9. Sperling, L.H. Polymeric Multicomponent Materials; John Wiley & Sons: New York, **1997**.
10. Sukhishvili, S. A.; Granic, S. *Phys. Rev. Lett.* **1998**, 80, 3646.
11. Schmitz, K. S. *Macromolecules* **2000**, 33, 2284.
12. Bicerano, J. *Prediction of Polymer Properties*; Marcel Dekker Inc., **2002**.
13. Chumanov, G.; Sokolov, K.; Cotton, T.M. *J. Phys. Chem.* **1996**, 100, 5166.
14. Xeu, G.; Dong, J. *Polymer* **1993**, 33, 643.
15. Lazarides, A. A.; Shatz, G. C. *J. Chem. Phys.* **2000**, 112, 2987.
16. Storhoff, J. J.; Lazarides, A. A.; Mucic, R. C.; Mirkin, C. A.; Letsinger, R. L.; Shatz, G. C. *J. Am. Chem. Soc.* **2000**, 122, 4640.
17. Kralchevsky, P. A.; Denkov, N. D. *Current Opinion in Colloidal & Interface Science* **2001**, 6, 383.
18. Coulon, G.; Russell, T.P.; Deline, V. R.; Green, P. F. *Macromolecules* **1989**, 22, 2581.
19. Rusell, T. P.; Coulon, G.; Deline, V.R.; Miller, D.C. *Macromolecules* **1989**, 22, 4600.
20. Kunz, M. S.; Shull, K. R.; Kellock, A. J. *J. Coll. Int. Sci.* **1993**, 156, 240.

Indexes

Author Index

Subject Index

A

Acrylamide (AM). *See* Latex film formation

Acrylic acid (AA). *See* Latex film formation

Aggregation
silver particles undergoing surface, 159–160
sodium dodecyl sulfate following desorption, 46

Ag nanoparticles. *See* Nanocomposite films with silver nanoparticles

Atomic force microscopy (AFM)
silane coated Kapton immersed in Au solutions, 143, 144*f*, 145*f*
surfactant transport during film drying, 54

Au film. *See* Conductive Au film on flexible polymer substrates

B

Biologically active phospholipids
colloidal particle shapes, 110, 112*f*
See also Colloidal particles

Block copolymer. *See* Nanocomposite films with silver nanoparticles

Buoy-anchor assembly, diblock copolymer, 162, 165

n-Butyl acrylate (BA). *See* Colloidal particles; Latex film formation

C

Cell size
time evolution of, vs. time for semi-crystalline copolymer rubber, 130*f*
See also Two-dimensional continuous network skeleton

Coalescing aid, latex film formation, 78, 80*f*, 81*f*

Cocklebur morphologies, 1,2-bis(10,12-tricosadiynoyl)-*sn*-glycero-3-phosphocholine (DCPC), 113, 115

Colloidal particles
aqueous polymerization of methyl methacrylate/*n*-butyl acrylate/heptadeca fluoro decyl methacrylate (MMA/nBA/FMA), 110, 111*f*
atomic force microscopy (AFM) and internal reflection infrared (IRIR) at film-air (F-A) interface of sodium dioctyl sulfosuccinate/1-myristoyl-2-hydroxyl-*sn*-glycero phosphocholine (SDOSS/MHPC) stabilized pMMA/nBA films, 117*f*
1,2-bis(10,12-tricosadiynoyl)-*sn*-glycerol-3-phosphocholine (DC$_{8,9}$PC) polymerizable phospholipid, 113, 114*f*
coalescence for dual tail phospholipids, 118, 119

171

Bestsellers from ACS Books

The ACS Style Guide: A Manual for Authors and Editors (2nd Edition)
Edited by Janet S. Dodd
470 pp; clothbound ISBN 0–8412–3461–2; paperback ISBN 0–8412–3462–0

Reagent Chemicals: Specifications and Procedures: Tenth Edition
By ACS Committee on Analytical Reagents
816 pp; clothbound ISBN 0–8412–3945–2

Advances in Arsenic Research: Integration of Experimental and Observational Studies and Implications for Mitigation
Edited by Peggy A. O'Day, Dimitrios Vlassopoulos, Xiaoguang Meng, and Liane G. Benning
446 pp; clothbound ISBN 0–8412–3913–4

Chemical Activities (student and teacher editions)
By Christie L. Borgford and Lee R. Summerlin
330 pp; spiralbound ISBN 0–8412–1417–4; teacher edition,
ISBN 0–8412–1416–6

Chemical Demonstrations: A Sourcebook for Teachers, Volumes 1 and 2,
Second Edition
Volume 1 by Lee R. Summerlin and James L. Ealy, Jr.
198 pp; spiralbound ISBN 0–8412–1481–6
Volume 2 by Lee R. Summerlin, Christie L. Borgford, and Julie B. Ealy
234 pp; spiralbound ISBN 0–8412–1535–9

The Internet: A Guide for Chemists
Edited by Steven M. Bachrach
360 pp; clothbound ISBN 0–8412–3223–7; paperback ISBN 0–8412–3224–5

Laboratory Waste Management: A Guidebook
ACS Task Force on Laboratory Waste Management
250 pp; clothbound ISBN 0–8412–2735–7; paperback ISBN 0–8412–2849–3

Metal-Containing and Metallosupramolecular Polymers and Materials
Edited by Ulrich S. Schubert, George R. Newkome, and Ian Manners
598 pp; clothbound ISBN 0–8412–3929–0

For further information contact:
Order Department
Oxford University Press
2001 Evans Road
Cary, NC 27513
Phone: 1-800-445-9714 or 919-677-0977

More Best Sellers from ACS Books

Microwave-Enhanced Chemistry: Fundamentals, Sample Preparation, and Applications
Edited by H. M. (Skip) Kingston and Stephen J. Haswell
800 pp; clothbound ISBN 0–8412–3375–6

Fire and Polymers IV: Materials and Concepts for Hazard Prevention
Edited by Charles A. Wilkie and Gordon L. Nelson
436 pp; clothbound ISBN 0–8412–3948–7

Ionic Liquids as Green Solvents: Progress and Prospects
Edited by Robin D. Rogers and Kenneth R. Seddon
614 pp; clothbound ISBN 0–8412–3856–1

Fermentation Biotechnology
Edited by Badal C. Saha
300 pp; clothbound ISBN 0–8412–3845–6

Chemometrics and Chemoinformatics
Edited by Barry K. Lavinex
216 pp; casebound ISBN 0–8412–3858–8

Polymeric Drug Delivery I: Particulate Drug Carriers
Edited by Sönke Svenson
352 pp; clothbound ISBN 0–8412–3918–5

Polymeric Drug Delivery II: Polymeric Matrices and Drug Particle Engineering
Edited by Sönke Svenson
390 pp; clothbound ISBN 0–8412–3976–2

Food Lipids: Chemistry, Flavor, and Texture
Edited by Fereidoon Shahidi and Hugo Weenen
248 pp; clothbound ISBN 978–0–8412–3896–1

Herbs: Challenges in Chemistry and Biology
Edited by Mingfu Wang, Shengmin Sang, Lucy Sun Hwang, and Chi-Tang Ho
384 pp; clothbound ISBN 978–0–8412–3930–2

For further information contact:
Order Department
Oxford University Press
2001 Evans Road
Cary, NC 27513
Phone: 1-800-445-9714 or 919-677-0977

Highlights from ACS Books